Springer Series in

CHEMICAL PHYSICS 98

Springer Series in

CHEMICAL PHYSICS

The purpose of this series is to provide comprehensive up-to-date monographs in both well established disciplines and emerging research areas within the broad fields of chemical physics and physical chemistry. The books deal with both fundamental science and applications, and may have either a theoretical or an experimental emphasis. They are aimed primarily at researchers and graduate students in chemical physics and related fields.

Please view available titles in *Springer Series in Chemical Physics* on series homepage http://www.springer.com/series/676

Kaoru Yamanouchi
Antonio Giulietti
Kenneth Ledingham
Editors

Progress in Ultrafast Intense Laser Science

Volume V

With 114 Figures

Springer

Editors

Professor Kaoru Yamanouchi
The University of Tokyo
Department of Chemistry
Hongo 7-3-1, 113-0033 Tokyo, Japan
E-Mail: kaoru@chem.s.u-tokyo.ac.jp

Professor Kenneth Ledingham
University of Strathclyde
Department of Physics
Glasgow, G4 0NG, United Kingdom
E-Mail: k.ledingham@phys.strath.ac.uk

Dr. Antonio Giulietti
Istituto per i Processi Chimico-Fisici
Intense Laser Irradiation Laboratory
1 Via G. Moruzzi, 56124 Pisa, Italy
E-Mail: tonino@ipcf.cnr.it

Springer Series in Chemical Physics ISSN 0172-6218
ISBN 978-3-642-03824-2 e-ISBN 978-3-642-03825-9
DOI 10.1007/978-3-642-03825-9
Springer Heidelberg Dordrecht London New York

Library of Congress Control Number: 2009938023

Cover design: SPi Publisher Services

Printed on acid-free paper

Springer is a part of Springer Science+Business Media (www.springer.com)

Preface

We are pleased to present the fifth volume of Progress in Ultrafast Intense Laser Science. As the frontiers of ultrafast intense laser science rapidly expand ever outward, there continues to be a growing demand for an introduction to this interdisciplinary research field that is at once widely accessible and capable of delivering cutting-edge developments. Our series aims to respond to this call by providing a compilation of concise review-style articles written by researchers at the forefront of this research field, so that researchers with different backgrounds as well as graduate students can easily grasp the essential aspects.

As in the previous volumes of PUILS, each chapter of this book begins with an introductory part, in which a clear and concise overview of the topic and its significance is given, and moves onto a description of the authors' most recent research results. All the chapters are peer-reviewed. The articles of this fifth volume cover a diverse range of the interdisciplinary research field, and the topics may be grouped into three categories: coherent responses of gaseous and condensed matter to ultrashort intense laser pulses (Chaps. 1–4), propagation of intense laser pulses (Chaps. 5, 6), and laser-plasma interaction and its applications (Chaps. 7–10).

From the third volume, the PUILS series has been edited in liaison with the activities of Center for Ultrafast Intense Laser Science in The University of Tokyo, and JILS (Japan Intense Light Field Science Society), the latter of which has also been responsible for sponsoring the series and making the regular publication of its volumes possible. From the present volume, the Consortium on Education and Research on Advanced Laser Science, the University of Tokyo, joins this publication activity as one of the sponsoring programs. The series has also collaborated since its inception with the annual symposium series of ISUILS (http://www.isuils.jp), which is designed to stimulate interdisciplinary discussion at the forefront of ultrafast intense laser science.

We would like to take this opportunity to thank all the authors who have kindly contributed to the PUILS series by describing their most recent work at the frontiers of ultrafast intense laser science. We also thank the reviewers who have read the submitted manuscripts carefully. One of the co-editors (KY) thanks Ms. Maki Oyamada and Ms. Chie Sakuta for their help with the editing processes. Last but not least, our gratitude goes out to Dr. Claus Ascheron, Physics Editor of Springer Verlag at Heidelberg, for his kind support.

We hope this volume will convey the excitement of Ultrafast Intense Laser Science to the readers, and stimulate interdisciplinary interactions among researchers, thus paving the way to explorations of new frontiers.

Tokyo,
Pisa,
Glasgow,
November 2009

Kaoru Yamanouchi
Antonio Giulietti
Kenneth Ledingham

Contents

List of Contributors

Tiberio Ceccotti
CEA, IRAMIS, SPAM
F-91191 Gif-sur-Yvette
France
tiberio.ceccotti@cea.fr

Aditya K. Dharmadhikari
Tata Institute of
Fundamental Research
1 Homi Bhabha Road
Mumbai 400 005
India
aditya@tifr.res.in

Li Fang
Department of Physics
University of Connecticut
Storrs
CT 06269
USA

Andrea Gamucci
Intense Laser Irradiation
Laboratory (ILIL),
Consiglio Nazionale delle
Ricerche (CNR),
IPCF - CNR Campus
Via G. Moruzzi, 1
56124 Pisa
Italy
and
Sezione di Pisa
Istituto Nazionale di
Fisica Nucleare
Italy

George N. Gibson
Department of Physics
University of Connecticut
Storrs
CT 06269
USA
gibson@phys.uconn.edu

Antonio Giulietti
Intense Laser Irradiation
Laboratory (ILIL),
Consiglio Nazionale delle
Ricerche (CNR),
IPCF - CNR Campus
Via G. Moruzzi, 1
56124 Pisa
Italy
and
Sezione di Pisa
Istituto Nazionale di
Fisica Nucleare
Italy

Danilo Giulietti
Dipartimento di Fisica
Università di Pisa
Italy
and
Intense Laser
Irradiation Laboratory
IPCF, Consiglio
Nazionale delle Ricerche
Pisa
Italy
and
Sezione di Pisa
Istituto Nazionale di
Fisica Nucleare
Italy
danilo.giulietti@df.unipi.it

Leonida A. Gizzi
Intense Laser Irradiation
Laboratory (ILIL),
Consiglio Nazionale delle
Ricerche (CNR),
IPCF - CNR Campus
Via G. Moruzzi, 1
56124 Pisa
Italy
and
Sezione di Pisa and
Laboratori Nazionali
di Frascati, Istituto
Nazionale di Fisica
Nucleare
Italy
la.gizzi@ipcf.cnr.it

Kunie Ishioka
National Institute for
Materials Science
Tsukuba, 305-0047
Japan
ishioka.kunie@nims.go.jp

Tsuneto Kanai
Laser Technology Laboratory
RIKEN, 2-1 Hirosawa
Wako-shi, Saitama
351-0198
Japan
tkanai@riken.jp

Jérôme Kasparian
Teramobile, GAP
Université de Genève
20 rue de l'École de Médecine
CH-1211 Genève 4
Switzerland
jerome.kasparian@unige.ch

Luca Labate
Intense Laser Irradiation
Laboratory (ILIL),
Consiglio Nazionale delle
Ricerche (CNR),
IPCF - CNR Campus,
Via G. Moruzzi, 1
56124 Pisa
Italy
and
Sezione di Pisa and
Laboratori Nazionali
di Frascati, Istituto
Nazionale di Fisica
Nucleare
Italy

Anna Lévy
CEA, IRAMIS, SPAM
F-91191 Gif-sur-Yvette
France

Philippe Martin
CEA, IRAMIS, SPAM
F-91191 Gif-sur-Yvette
France

Deepak Mathur
Tata Institute of
Fundamental Research
1 Homi Bhabha Road
Mumbai 400 005
India
atmol1@tifr.res.in

Katsumi Midorikawa
Laser Technology Laboratory
RIKEN, 2-1 Hirosawa
Wako-shi, Saitama
351-0198
Japan

Oleg V. Misochko
Institute of Solid State Physics
Russian Academy of Sciences
Chernogolovka
142432, Moscow region
Russia

Bradley Moser
Department of Physics
University of Connecticut
Storrs
CT 06269
USA

Yasuo Nabekawa
Laser Technology Laboratory
RIKEN, 2-1 Hirosawa
Wako-shi, Saitama
351-0198
Japan

Eiji J. Takahashi
Laser Technology Laboratory
RIKEN, 2-1 Hirosawa
Wako-shi, Saitama
351-0198
Japan

Jean-Pierre Wolf
Teramobile, GAP, Université de
Genève, 20 rue de l'École de
Médecine, CH-1211
Genève 4
Switzerland

1

Vibrational and Electronic Excitation of Molecules by Short-Pulse Strong Laser Fields

George N. Gibson, Li Fang, and Bradley Moser

Summary. For weak electromagnetic fields, vibrational and electronic excitation of molecules generally occurs through resonant one-photon interactions or two-photon Raman-type transitions. Short-pulse strong laser fields open up a variety of new excitation schemes, which are characterized by non-resonant or high-order resonant interactions. The study of vibrational and electronic excitation of molecules by strong laser fields is important for a number of reasons: (1) It reveals new features of the strong field interaction with molecules, such as R-dependent ionization and inner orbital ionization; (2) It informs other studies such as high-harmonic generation and quantum tomography. In particular, it probes various assumptions that are made, such as only one orbital is active and the ionization rate is a maximum along the internuclear axis; (3) It allows for the spectroscopy of previously inaccessible states of the molecules, such as the charge resonant state of I_2^{2+} and the excited states of H_2^+; (4) It may lead to novel methods for creating population inversions in the vaccum–ultraviolet; (5) It may lead to new methods for coherent control of thermal ensembles. In this paper, we present an overview of these strong field excitation schemes and give examples for many of them.

1.1 Introduction

The interaction of even simple diatomic molecules with strong laser fields is considerably more complicated than the interaction with atoms. In atoms, nearly all of the observed phenomena can be explained with a simple three-step model [1], at least in the tunneling regime: (1) The laser field releases the least bound electron through tunneling ionization; (2) the free electron evolves in the laser field; and (3) under certain conditions, the electron can return to the vicinity of the ion core, and either collisionally ionize a second electron [2], scatter off the core and gain additional kinetic energy [3], or recombine with the core and produce a harmonic photon [4].

The interaction with molecules is much more complex because, first, the sequence of events described above now depends on the additional degrees of freedom in the molecule, including angular orientation [5] and internuclear separation [6] and, second, new processes occur, such as dissociation [7] and vibrational [8] and electronic excitation [9]. Much of the work on molecules in strong laser fields has focused on the dependence of strong field phenomena, such as ionization [5, 6] and high-harmonic generation [10], on angular orientation and internuclear separation. The study of the kinetic energy release (KER) of the ion fragments following ionization and dissociation has also received considerable attention [11], as have attempts to control molecules [12], in particular their angular alignment and orientation [13].

The vibrational and electronic excitation of molecules has received less attention over the years, but the understanding of excitation processes is important for a number of reasons. Thus, in this paper, we will review the various mechanisms leading to vibrational and/or electronic excitation of molecules in strong laser fields. As we do this, it will become clear that exploring these mechanisms (1) reveals new features of the strong field interaction; (2) has important consequences for the new field of quantum tomography; (3) opens up the possibility of transient spectroscopy on previously inaccessible states of molecules; and, (4) may lead to techniques for creating population inversions in the vacuum–ultraviolet (VUV) spectral region.

The paper is organized as follows: we first discuss vibrational excitation through various mechanisms, including ionization, R-dependent depletion, and bond-softening. We then present evidence for electronic excitation and consider multiphoton excitation, inner orbital ionization, and excitation through recollisions. Several applications of these interactions are presented, followed by our conclusions.

1.2 Vibrational Excitation

Many experiments on molecules in strong laser fields are performed with single laser pulses. With this, ions [11], electrons [14], and photons [15] produced by the laser interaction can be measured as a function of intensity, wavelength, pulse duration, etc. However, under these conditions, vibrational motion induced by the interaction will not be seen directly, as the distinguishing feature of the vibrational motion is its time dependence. The situation changes completely with a simple two-pulse pump-probe arrangement. In this case, even basic ion time-of-flight (TOF) spectroscopy shows dramatic vibrational signals on many dissociation channels [16, 17] (see Fig. 1.1). The vibrational amplitude and frequency depend on many factors, indicating a variety of excitation pathways.

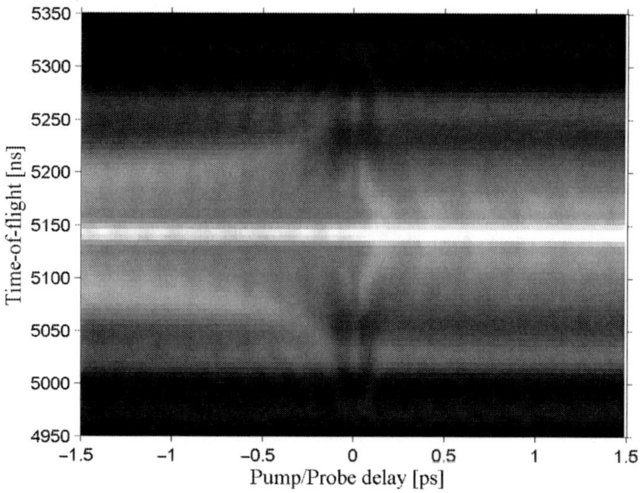

Fig. 1.1. TOF data for I_2 as a function of time delay in a pump-probe experiment showing a variety of vibrational signals in the I^+ signal. The horizontal band at 5,145 ns corresponds to metastable I_2^{2+}. The pair of bands at 5,020 and 5,270 ns corresponds to the $I^+ + I^+$ dissociation channel, while the signal in between these bands corresponds to $I^+ + I$. Pump and probe are both 25 fs, 800 nm pulses

1.2.1 Vibrational Excitation Through Ionization

The simplest way to produce vibrational motion with a non-resonant short pulse laser field is to ionize a molecule where the ground state of the molecular ion has a potential well significantly offset from the neutral ground state in internuclear separation R. H_2 serves as a clear, well-studied example (see Fig. 1.2 [18]). With short pulse ionization, the vibrational wavefunction in the electronic ground state of the neutral molecule is projected more or less vertically into the ground state of the ion [19]. The offset of these potential curves in H_2 is quite large – the most probable vibrational level populated in H_2^+ is $v = 2$ and there is significant population up to $v = 8$. However, just as important, a well-defined wavepacket is produced that evolves coherently. While this vibrational motion was originally deduced from single-pulse TOF measurements to explain the dissociation of H_2^+ into $H + H^+$ [20], this motion has been directly observed in pump-probe measurements with 8 fs laser pulses [21]. N_2 [22] and I_2 [23] are often studied molecules, but the offsets of the potential curves in the neutral and ionic states are rather small, suggesting that little vibration motion will be induced through ionization (see Fig. 1.3 [24, 25]). Indeed, we have found no evidence for vibrational motion in the ground state of either N_2^+ or I_2^+. However, another possibility exists: many excited states of molecules have large equilibrium separations R_{eq}. Ionization to one of these states would lead to large amplitude motion, as in H_2^+. Such vibrational motion was seen following ionization of I_2 to I_2^+ and the vibrational

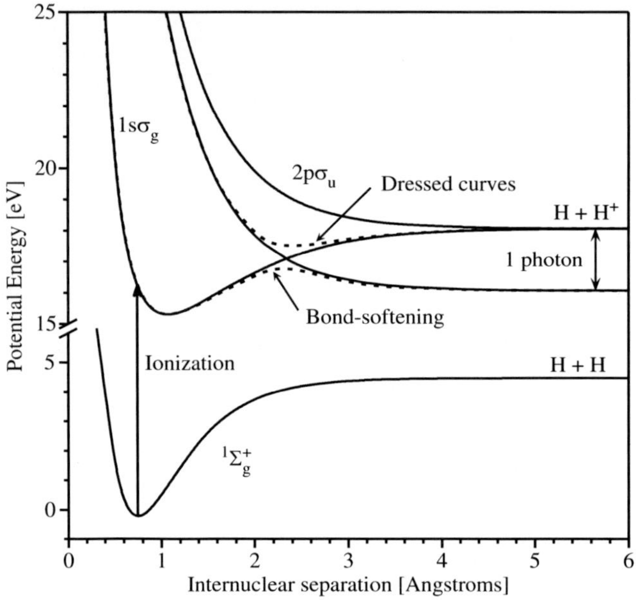

Fig. 1.2. Potential energy curves of H_2 and H_2^+ showing ionization and dressed states in a laser field. The dressed curves lead to bond softening and a distortion of the potential curve of the ground state of the ion, as will be discussed in Sect. 1.2.3

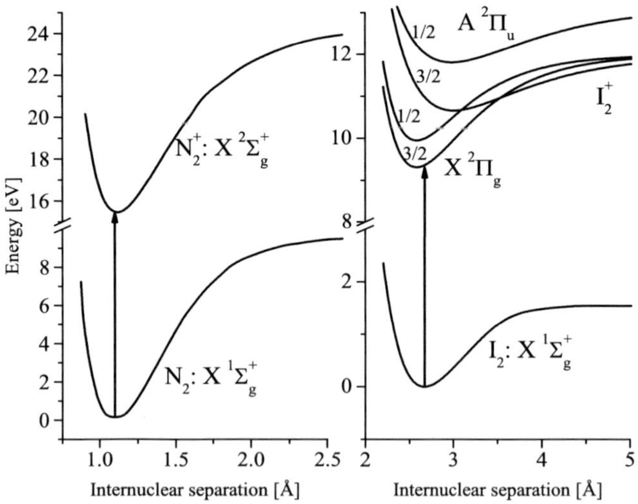

Fig. 1.3. Selected potential energy curves of N_2, N_2^+, I_2, and I_2^+

signature of the motion identified the state as the A state [17]. This is an interesting example of the interplay between vibrational and electronic excitation. In normal TOF spectroscopy, ionization of I_2 produces a signal at the I_2^+ TOF peak. However, there is no information about whether the I_2^+ is in the ground state or an excited state. Given the success of tunneling ionization theories in explaining strong field ionization [26], the natural assumption would be that the I_2^+ molecules are in the ground state. Using vibrational spectroscopy, we now know that I_2^+ can, at least, be produced in the A state. This is interesting in that the A state is formed by the removal of an inner orbital electron, the π_u rather than the π_g [25]. This will be discussed more in Sect. 1.3.3.

1.2.2 Vibrational Excitation Through *R*-Dependent Depletion

The first issue involved in studying vibrational excitation is identifying the electronic state of the molecule. While the first excitation mechanism discussed above immediately suggests electronic states of the molecular ion, some of the vibrations in Fig. 1.1 appeared to come from the ground state of the neutral molecule. While it is clear how ionization to an offset state of the ion produces large amplitude vibrations, producing large vibrations in the ground state is not so easy. However, ground state vibrations in H_2 were observed in [27] and they attributed them to "*R*-selective ionization," which they dubbed "Lochfrass" (LF) or "hole eating." This is a purely quantum mechanical effect that can produce vibrational excitation through non-resonant ionization. All that is required is for the ionization rate from a neutral molecule to have a strong dependence on R. If a large fraction of the ground state wavefunction is ionized, the remaining or unionized portion of the wavefunction will no longer have the correct shape to be in the ground vibrational state. Thus, it must be in a superposition of vibrational states, which will then evolve in time to produce vibrational motion. It is quite common in molecules to have an increasing ionization rate as a function of increasing R, at least for R close to R_{eq}. In this case, the part of the wavefunction at large R will be preferentially ionized, producing a "hole" in the wavefunction, as shown schematically in Fig. 1.4.

While the idea of LF explained the H_2 data quite well [28], we were surprised by the magnitude of the oscillations in our I_2 data [16], as, unlike H_2, I_2 is not vibrationally cold at room temperature – the conditions for our experiment. Generally, thermal motion is detrimental to observing coherent motion. Thus, we took a long time scale run to get a more accurate measurement of the frequency of the vibrations, shown in Fig. 1.5. These data also exhibit a vibrational revival, from which the anharmonicity of the potential well can be determined. Indeed, the vibrational frequency accurately matched that of the ground state.

Simulations of LF in a model molecule similar to I_2 confirm that LF can produce coherent vibrations in hot molecules and, in fact, the amplitude of the motion *increases* with increasing temperature. This is unique in the field of

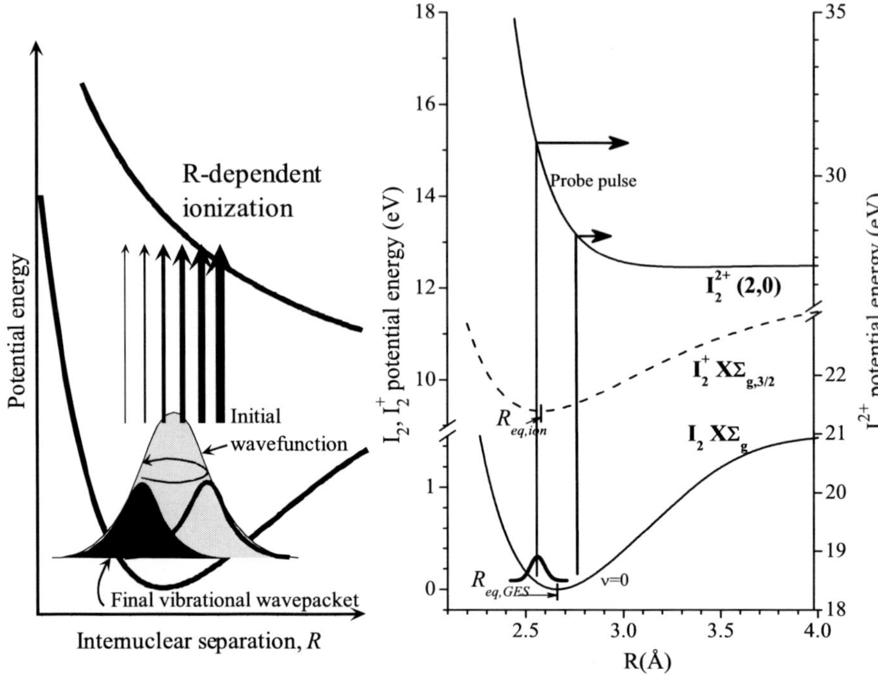

Fig. 1.4. Vibrational motion induced by Lochfrass as well as the detection scheme using a repulsive curve of I_2^{2+} [16]

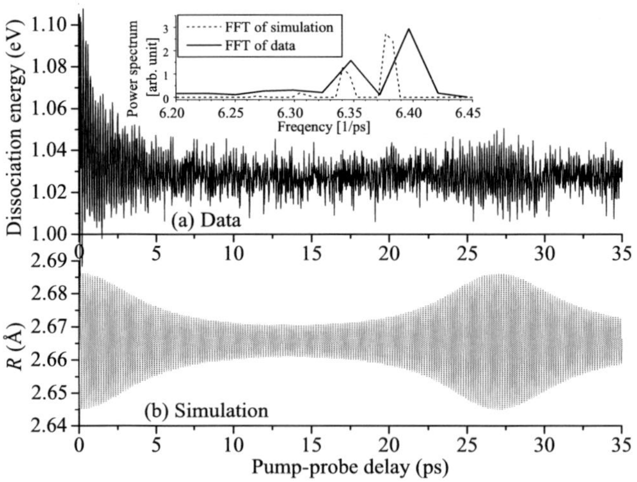

Fig. 1.5. Revivals in I_2 following vibrational excitation with Lochfrass. (**a**) Data from experiments similar to [16]. (**b**) Simulations of vibrational oscillations in the I_2 ground electronic state

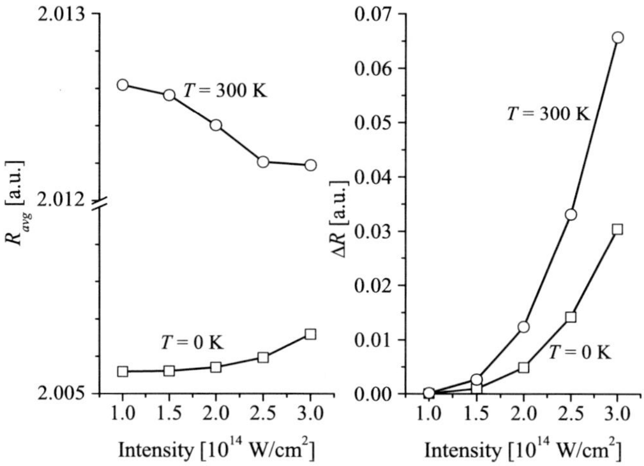

Fig. 1.6. Results from simulations showing the simultaneous increase in vibrational amplitude ΔR and decrease in R_{avg} for Lochfrass in thermal molecules. The pulse duration is 25 fs and the calculations assume tunneling ionization (i.e., no dependence on wavelength)

coherent control and may lead to better coherent control at room temperature. So, although thermal ensembles generally are detrimental to creating coherent motion, the LF mechanism actually works better in hot molecules than in cold.

Another remarkable aspect of LF, seen in the simulations, is that it can actually vibrationally cool the molecules. In particular, this means that the average internuclear separation R_{avg} should decrease as a function of increasing intensity (Fig. 1.6). This is exactly what we observe experimentally in Fig. 1.7, even as the amplitude of the vibrations is increasing [29].

1.2.3 Vibrational Excitation Through Bond-Softening

In [27], while trying to understand the mechanism for producing ground state vibrations in H_2, the authors considered bond-softening (BS) in addition to LF. Bond-softening was first proposed to explain the KER of H^+ fragments following the strong field ionization of H_2 [20]. Bond-softening occurs when a bound potential energy curve is strongly coupled by the laser field to a higher repulsive curve (see Fig. 1.2). The bound curve takes on repulsive character, leading to a "softening" of the bond. Conversely, the repulsive curve can take on sufficient bound character to have a (transient) bound state. This has been observed and called "bond-hardening" [30]. In any case, it is possible that this coupling can produce a sufficiently large distortion of the ground state potential curve in H_2 to induce vibrational motion. In a sense, this is just a time-dependent description of stimulated Raman scattering. However, simulations show that BS produces vibrational motion with a different phase than LF (see Fig. 1.8) and operates in a different intensity range [28], allowing

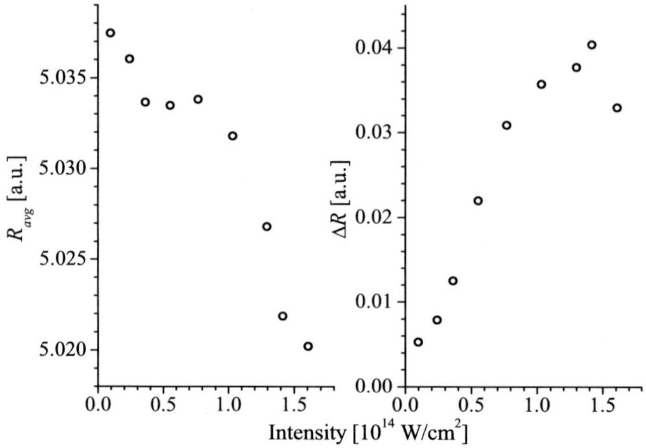

Fig. 1.7. Experimental results for ΔR and R_{avg} in room-temperature I_2, which qualitatively agree with Lochfrass

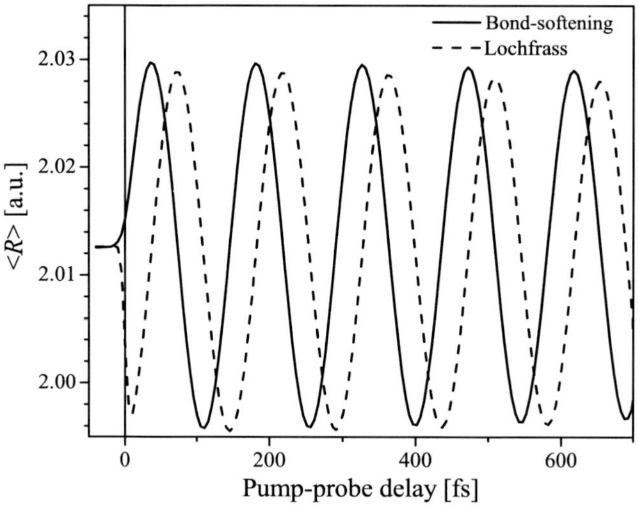

Fig. 1.8. Ground state vibrations resulting from Lochfrass and bond-softening showing different phases of the motion. The pulse duration is 25 fs. The intensity is $2.5 \cdot 10^{14}\,\mathrm{W\,cm^{-2}}$ for Lochfrass and $3.4 \cdot 10^{13}\,\mathrm{W\,cm^{-2}}$ for bondsoftening. These were chosen to make the amplitude of the motion the same for the two processes. The vibrational phase does not depend on intensity. The wavelength for the bond-softening calculation is 400 nm

the authors in [27] to identify LF rather than BS as the excitation mechanism. In a similar way, we were able to rule out BS in the experiments discussed above in Sect. 1.2.2. However, we also found from simulations that BS cannot lead to vibrational cooling of the molecule, as it is a non-dissipative reversible interaction. As we found experimentally that the interaction does lead to cooling, we could further rule out BS as the excitation mechanism. Nevertheless, under certain conditions, BS should induce vibrational motion.

1.3 Electronic Excitation

Excitation does not appear to play a very important role in the strong field interaction with atoms. Generally speaking, the excited states of the rare gases tend to lie rather high in energy and there is no effective means to excite these states. However, the situation in molecules is apparently quite different, as there are many examples of significant population ending up in excited states of molecules or their ions.

1.3.1 Ionization to Charge-Asymmetric Dissociation Curves

The simplest and first observation of the excitation of molecules by strong laser fields was the identification with TOF spectroscopy of the following dissociation channel: $I_2^{2+} \rightarrow I^{2+} + I^{0+}$ [31]. In this example, a short laser pulse doubly ionizes an I_2 molecule. The ground state manifold of I_2^{2+} consists of metastable potential curves and curves that dissociate to $I^{1+} + I^{1+}$, the symmetric dissociation channel. The dissociation limit of the excited state curves leading to the asymmetric channel, $I^{2+} + I^{0+}$, is 8.7 eV higher in energy than the limit of the symmetric channel [32] (see Fig. 1.9). Thus, the simple observation of the asymmetric channel means that excited molecular states are populated through strong field ionization. This property is not unique to I_2^{2+}.

Indeed, all even charge states of I_2 and N_2 that we have studied have both a symmetric and an asymmetric dissociation channel, up to I_2^{12+} [33] and N_2^{6+} [34]. Such dissociation channels can be unambiguously identified through correlations between, for example, the $N^{2+} + N^{4+}$ ion signals, as shown in Fig. 1.10. This particular asymmetric channel in N_2^{6+} is 30 eV higher in energy than the symmetric channel. At 800 nm, this is equivalent to absorbing about 20 photons. While the mechanism for this excitation is still unclear, the results in the next section may provide a clue.

Finally, it has been observed through pump-probe experiments [9] that the I^{2+} fragment following asymmetric dissociation of I_2^{2+} can itself be in an excited state. This also represents a very high level of electronic excitation.

1.3.2 Resonant High-Order Multiphoton Excitation

The asymmetric dissociation channels discussed in the previous section are closely tied to the charge-resonant states first introduced by Muliken in 1939 [35]. These states have no counterpart in atoms and occur whenever there is a difference in charge state in the dissociation limit of a diatomic molecule. They

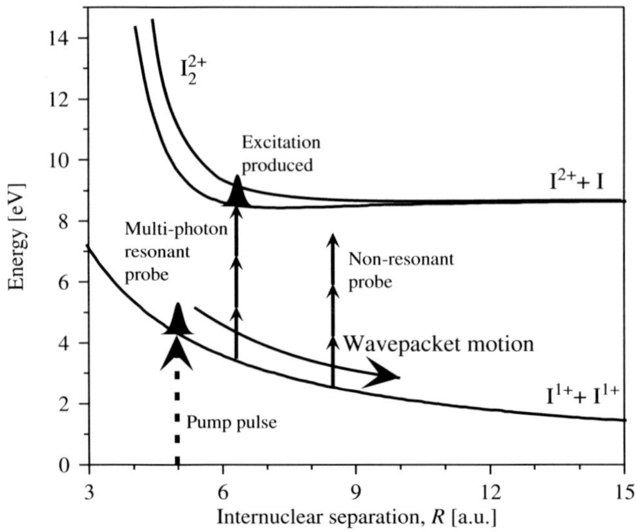

Fig. 1.9. Potential energy curves of I_2^{2+} showing the symmetric and asymmetric dissociation channels. Also shown is an example of transient pump-probe multiphoton spectroscopy, discussed in Sect. 1.3.2

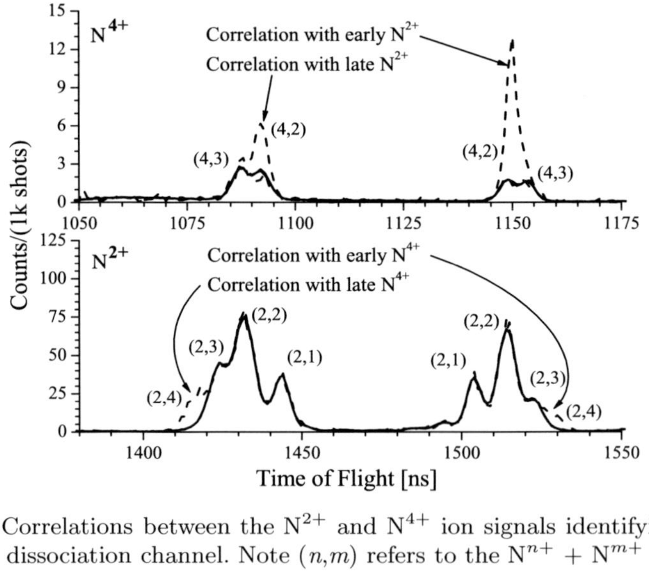

Fig. 1.10. Correlations between the N^{2+} and N^{4+} ion signals identifying the N_2^{6+} asymmetric dissociation channel. Note (n,m) refers to the N^{n+} + N^{m+} dissociation channel

occur in gerade and ungerade pairs that have two important properties: they asymptotically become degenerate at large R and they have a dipole coupling that grows linearly with R [36]. An example of this molecular system is shown in Fig. 1.11.

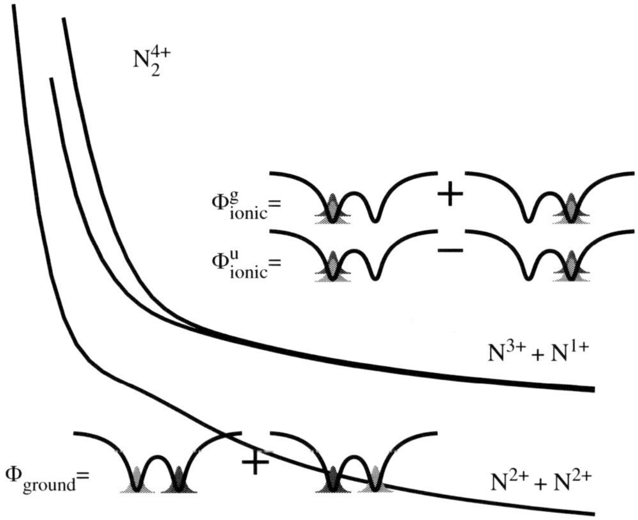

Fig. 1.11. Ground state and excited charge resonant states of a model doubly-charged diatomic molecule. The electronic configurations of the states are shown schematically in a 1-D double-well potential

It has been noted that these states have an interesting property for excitation [36,37]. If we consider a three-level system consisting of a ground state and two excited states that form a charge-resonant pair, then there is a strong multiphoton coupling between these states (see Fig. 1.12). Indeed, from solutions to the three-level system and full 1D 2-e quantum calculations, it has been shown that up to 12-photon resonant transitions are possible. Such transitions are simply not possible in a two-level system. The strong multiphoton coupling comes from the large dipole coupling between the charge–resonant states. They are strongly modulated by an external field, which essentially adds sidebands and was dubbed a "Floquet ladder" of states. These sidebands produce the multiphoton transitions. Such transitions are difficult to realize in a two-level system as the multiphoton coupling is weak and, at high field strengths, the ac-Stark shift becomes large. In the three-level system, the multiphoton coupling comes from the interaction of the pair of charge–resonant states, which also experience a Stark shift, but the crucial difference here is that it is a linear Stark shift, as the states are degenerate. Although the Stark shift can be large, it averages to zero, resulting in no ac-Stark shift. In this way, the pair of excited states can stay in resonance with the ground state, even while being driven sufficiently hard to produce a high-order multiphoton transition.

Such a system was independently realized a few years later in a completely different field: driven quantum flux qubits [38]. Here, a superconducting loop can support a quantum unit of current in either direction around the loop. In an external dc magnetic field, the degeneracy of the two directions is lifted,

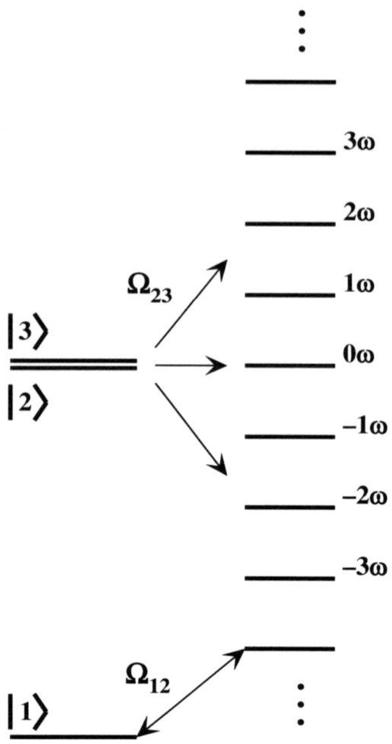

Fig. 1.12. The three-level system showing the creation of a Floquet ladder of states. Also shown is the weak one-photon coupling needed to make a transition to the Floquet ladder from the ground state. Ω_{12} and Ω_{23} are the Rabi frequencies between the states

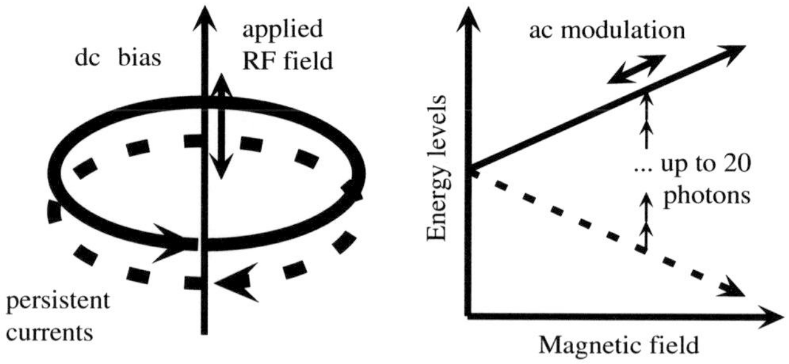

Fig. 1.13. Energy diagram for driven flux qubits that supports high-order multi-photon transitions

producing a two-level system. Transitions can then be made between these two levels with an external ac magnetic field (see Fig. 1.13). Resonant transitions at up to 20 times the driving frequency were seen. Although this appears, at first, to be like a normal two-level system, it is actually analogous to the three-level system described earlier. The reason is that the interaction of the two states with the external magnetic field is linear, as the states were originally degenerate. The ac magnetic field rides on top of the dc field and so the interaction is still linear. Again, the coupling is large but the ac Stark shift is zero.

As suggested earlier, homonuclear diatomic molecules contain this three-level structure and may explain the high level of excitation leading to the asymmetric dissociation channels. However, a more explicit demonstration was performed in a pump-probe experiment on I_2^{2+} [39]. As discussed earlier, the ground state of I_2^{2+} dissociates into $I^+ + I^+$ (referred to as the (1,1) channel), and so this channel reflects the ground state population. The excited charge-resonant states dissociate into $I^{2+} + I^{0+}$, or (2,0), which gives a measure of the excited state population (see Fig. 1.9). In this experiment, the pump pulse doubly ionizes I_2 into the ground state of I_2^{2+}, which then begins to dissociate. Then, a weaker probe pulse is applied to resonantly transfer population to the excited (2,0) state. For a fixed probe wavelength, the transition will not, in general, be resonant and no transfer should occur. However, as the molecule dissociates, there can be particular R's where the laser does come into (multiphoton) resonance with the (1,1) to (2,0) transition. If so, at particular time delays, there would be an increase in the (2,0) signal along with a decrease in the (1,1) signal. In fact, such a delay-dependent signal was seen, indicating an R-specific multiphoton transition. Unfortunately, the potential energy curves of I_2^{2+} are not well-known and, thus, we could not independently verify that the transition occurred at a resonance point. However, it does strongly suggest that the charge–resonant states do play an important role in the strong field interaction.

While the charge symmetric/asymmetric dissociation channels occur in even-charged molecular ions, a similar system occurs in odd-charged molecules, for example, H_2^+ [18]. Here, the two lowest states, the $1s\sigma_g$ and $2p\sigma_u$, form a pair that are degenerate at large R and have a dipole coupling that grows linearly with R. Using the same reasoning as earlier, these states can couple to excited states through resonant multiphoton interactions (see Fig. 1.14). Figure 1.15 shows simulations of a 1D model of H_2^+ in an external field. The internuclear separation is treated quantum mechanically while the electronic excitation includes only three levels at a time: the $1s\sigma_g$, the $2p\sigma_u$, and one excited state leading to the $n=2$ dissociation limit. As is seen in the figure, high-order multiphoton transitions to excited states of H_2^+ are quite strong. Unlike I_2^{2+}, it is difficult to monitor the ground and excited state population through TOF spectroscopy, as all states lead to the same dissociation channel, $H^+ + H$. However, laser-induced fluorescence (LIF) selectively monitors the excited state population, in this case the Lyman-α (L_α) line at 121.6 nm [40]. For the two lowest levels, the H atom comes out in the ground state, while the excited states will leave the H atom in the $2s$ and $2p$ states. The $2p$ state will eventually decay through fluorescence. In fact, this L_α radiation was observed following strong field ionization and excitation of H_2. It was determined that the excitation was due to direct laser excitation rather than plasma processes by measuring a linear pressure dependence to the L_α signal and showing that the radiation could be quenched within 100 fs with a second laser pulse.

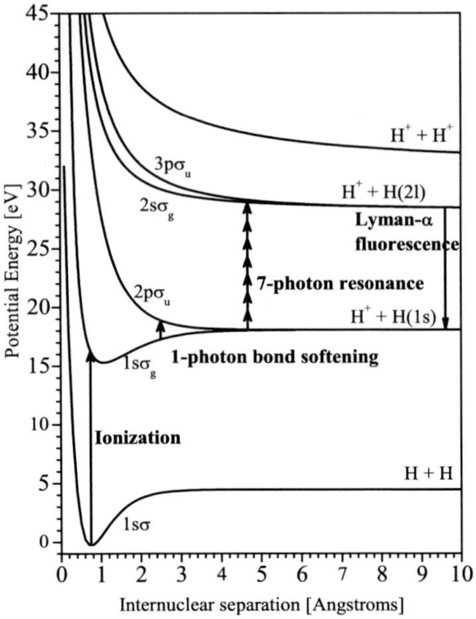

Fig. 1.14. Multiphoton excitation scheme in H_2^+ showing the laser-induced fluorescence detection method

Plasma excitation would occur on a much longer timescale, while fluorescence from prompt excitation of the $2p$ level in H could be readily quenched through the ionization of the $2p$ state with the second laser pulse.

As in iodine, although we found direct excitation through the strong field interaction, we could not be sure that it resulted from the resonant three-level model discussed earlier. However, a wavelength study provided strong evidence of resonant excitation. From Fig. 1.15, the best candidate for observing a resonant signature is the four-photon excitation of the $2p\pi_u$ state at around 495 nm. This sharp feature occurs because the $2p\pi_u$ potential curve closely follows the $1s\sigma_g$ and $2p\sigma_u$ states over a large range in R. Thus, the states are in resonance for a long time as the molecule dissociates. Figure 1.15 also shows the experimental L_α signal, from two independent scans, as a function of laser wavelength over the region where we expect to see the resonant feature. Indeed, a peak is observed, although the center is shifted from the expected value. This is probably due to ac Stark shifts from other levels that are not taken into account in our simulations. Again, this suggests that resonant high-order multiphoton transition do occur in diatomic molecules.

Finally, there is one other experiment that we know of that found a resonant signature in the strong field excitation of a molecule. In this experiment, the symmetric dissociation channel of O_2^{2+} was monitored in a high resolution coincidence ion spectrometer [41]. From the KER of the fragments, excitation to the $B^3\Pi_g$ state was identified, among other states. However, this particular

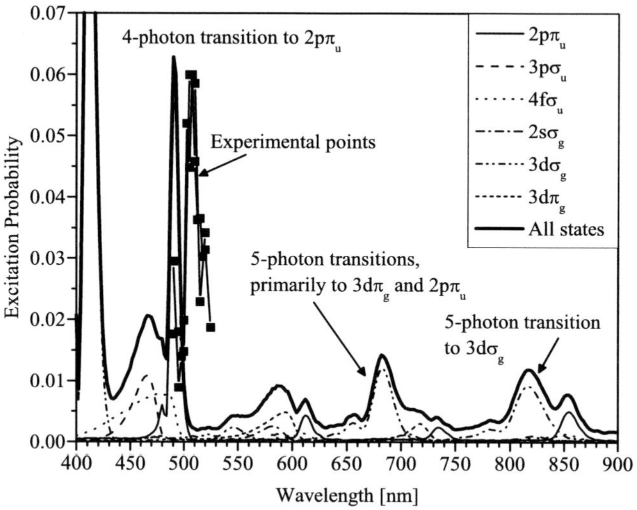

Fig. 1.15. Three-level calculations with full 1D vibrational motion of the excitation of H_2^+ as a function of wavelength. Also included are two independent sets of experimental data for wavelength scans from 490 to 520 nm

state showed a strong dependence on laser wavelength and, in fact, the excitation followed a resonant profile centered around 890 nm. It is not clear if this is due to the same three-level structure discussed above, but it does provide further evidence that resonant multiphoton excitation is important for molecules in strong laser fields.

1.3.3 Inner Orbital Ionization

Another excitation mechanism for molecules involves inner-orbital ionization. Although it is generally assumed that the least bound electron in an atom or molecule is ionized through tunneling induced by a strong laser field, this is not the case in molecules. If an electron besides the least bound one is ionized, the resulting ion will, by definition, be in an electronically excited state. In atoms, the inner shell electrons tend to lie rather deep in energy and it is unlikely that they can be ionized. However, in molecules, the binding energy of the inner orbital electrons varies greatly for different molecules. In N_2 [24] and I_2 [25], the higher lying inner orbitals are only a few electron-volts deeper than the least bound orbital. As discussed earlier, sometimes the excited states of a molecular ion can be identified through its vibrational frequency, as in the case of the A state of I_2^+ [17]. Thus, the observation of the A state in the TOF spectrum demonstrates an example of inner orbital ionization. While an inner orbital should have a smaller ionization rate than the outer orbital because of its greater binding energy, this factor may be overcome by the spatial structure of the orbital. Indeed, it was found in [17] that the inner

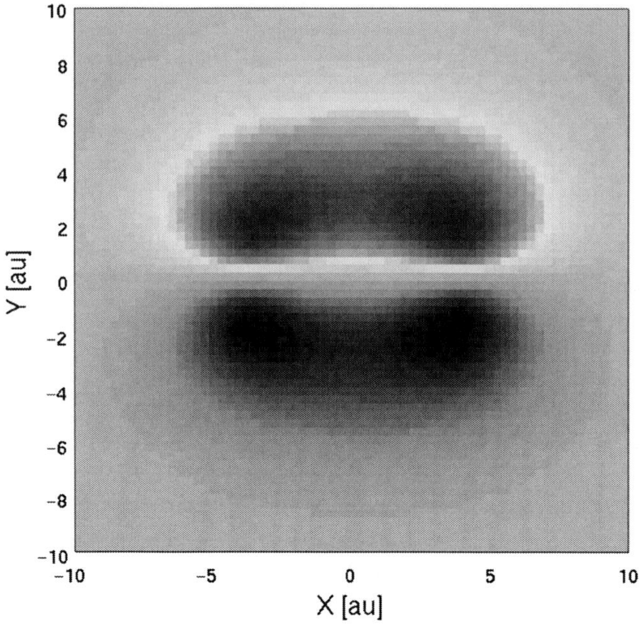

Fig. 1.16. Shape of the π_u orbital for a model diatomic molecule

π_u orbital was observed only with the laser polarization perpendicular to the molecular axis. Figure 1.16 shows the qualitative shape of a π_u orbital, which suggests that a field applied perpendicular to the molecular axis would be more effective in ionizing this orbital.

Older work using electron spectroscopy to study the strong field ionization of molecules showed directly the population of the three lowest lying states of N_2^+: the X, A, and B states, corresponding to the ionization of the $3\sigma_g$, the $1\pi_u$, and the $2\sigma_u$ orbitals [14]. More indirect work using VUV fluorescence identified the ionization of the significantly deeper $2\sigma_g$ orbital [42].

1.3.4 Excitation Through Recollision

Lastly, we mention one more excitation mechanism that has been observed in molecules. It is well-established that following strong field ionization in atoms and molecules, under certain conditions, the ionized electron can be driven back to the ion core where it can recombine to produce high-harmonic radiation, induce further ionization, or experience inelastic scattering. However, there is also the possibility of collisional excitation. Such excitation was observed in [43] in N_2 and O_2. In both molecules, one electron is tunnel ionized by the strong laser field. When the electron rescatters with the ion core, it can collisionally ionize and excite the molecular ion, creating either N_2^{2+} or O_2^{2+} in an excited state. When the double ion dissociates, its initial state can be determined from the KER of the atomic fragments. In this way, a number of excited states were identified in N_2 and O_2.

1.4 Applications

While excitation processes in strong laser fields have not received as much attention as ionization rates and high-harmonic generation, the study of excitation is an important aspect of our overall understanding of the behavior of molecules in strong laser fields.

1.4.1 Strong Field Interactions

Important aspects of the interaction of strong laser fields with molecules can be missed in standard TOF experiments, most notably the population of electronically excited states. However, by studying vibrational excitation, the frequency and dephasing of the vibrational motion can be used to identify the electronic state undergoing the vibrational motion. In some cases, this turns out to be a ground state, and in others, an excited state. Once we have identified an excited state, we are left with the question of how and why the state was populated by the strong field. In one example above (the I_2^+ A state discussed in Sect. 1.3.3), the excited state is formed by the removal of an inner orbital electron, in this case a π_u electron. This correlates with the measured angular dependence for the ionization to this state.

In the case of vibrational motion in the ground electronic state produced by LF, the study of the vibrational motion can provide information about how the ionization rate of the state depends on R. For I_2, certain theoretical calculations predicted that the ionization rate would decrease as R increased, at least near R_{eq} [28]. However, the experimental results showed the opposite trend.

1.4.2 Quantum Tomography

Quantum tomography refers to a technique by which the angular dependence of high-harmonic generation can be used to reconstruct the electron orbital of the least bound electron in a molecule. This technique has been used rather successfully to determine the $3\sigma_g$ orbital of N_2 [44]. However, one basic assumption is that only one orbital is involved in ionization in a strong field. As we have shown earlier, this assumption is not always true. In I_2, we find a significant amount of population in the excited A state, in addition to the ground state of I_2^+. This will become more likely with larger molecules as the orbital energies become closer together. Even in N_2, there have been observations of ionization to the excited states of N_2^+ [14].

The example of I_2 actually has a bigger problem. In quantum tomography, the angular dependence to the ionization rate must be taken into account for an accurate reconstruction. As we mentioned, ionization from inner orbitals will interfere with the reconstruction. In I_2, it turns out that the two problems are coupled. Ionization from the ground state is a maximum when the molecule is aligned with the laser polarization and a minimum when perpendicular. We have measured just the opposite dependence for ionization to the excited state: ionization of the inner orbital maximizes when the molecule is in the perpendicular orientation [17]. Thus, the admixture of ionization from the inner orbital changes as a function of angle. This will severely complicate any orbital reconstruction.

1.4.3 Molecular Spectroscopy

The identification of excited states in strong field interactions with molecules has lead to some novel forms of molecular spectroscopy, allowing previously inaccessible states to be studied. For the most part, this comes from the ability to do transient spectroscopy in the time domain with ultrashort pulses. But, the strong field interaction also allows for new population mechanisms.

As discussed earlier, the strong field ionization of I_2 can remove an inner-orbital electron, leaving the I_2^+ ion in the electronically excited A state. Moreover, a wavepacket is left in a rather high vibrational state, because the potential wells of the I_2 ground state and the A state are significantly offset. The wavepacket then evolves in time, probing a large range of internuclear separation. With a second ultrashort laser pulse, the A state can be further ionized to I_2^{2+}. This ionization is detected by monitoring the (2,0) dissociation channel. Any vibrational modulation on this signal comes from the vibrational characteristics of the A state. However, the KER of the (2,0) channel is determined by the potential energy curves of both the A state and the (2,0) state. As the A state is already known, the KER provides a direct measurement of the (2,0) state. This state is interesting as there is no long-range Coulomb repulsion. Moreover, there is a strong polarization attraction between the I^{2+} ion and the I atom. This raises the possibility of a bound potential well in the (2,0) curve. By measuring the KER of the (2,0) state as a function of time in a pump-probe experiment, we were able to determine that the (2,0) does, in fact, have a potential well with a depth of 60 meV, shown in Fig. 1.17 [17]. Interestingly, the outer turning point of the wavepacket in the A state is very close to the minimum energy of the (2,0) state, which means that we could actually populate the bound vibrational levels of the (2,0) curve. Indeed, at this point the (2,0) dissociation signal vanished, implying that we had created bound (2,0) that would no longer dissociate.

We have already discussed above the transient multiphoton spectroscopy of the (1,1) and (2,0) states in I_2^{2+} made possible by the strong multiphoton coupling from the ground state to the charge–resonant states [39]. By varying the laser wavelength of the probe pulse, it should be possible to map out the potential energy curves of high lying excited states.

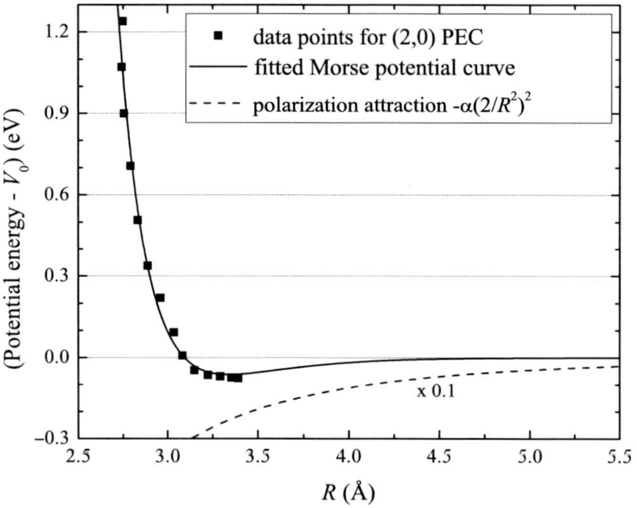

Fig. 1.17. Experimentally determined potential energy curve for the state leading to the (2,0) dissociation limit

Finally, it is interesting to note that there is virtually no experimental data on the excited states of H_2^+, even though this is the prototypical molecular system. The problem is that the excited states lie rather high in energy and are mostly unbound, making them difficult to access. However, we have found a resonant signature in the multiphoton excitation of the $2p\pi_u$ state, providing the first ever resonant multiphoton excitation of H_2^+ (see Fig. 1.15).

1.4.4 Population Inversions

One final application of understanding and controlling excitation in molecules by strong laser fields involves the creation of real population inversions in the VUV spectral region. Neutral molecules, for the most part, will not support inversions in the VUV spectral region, as the ionization potentials are too low. Thus, it is necessary to work with ionized molecules, which tend to dissociate rapidly. Nevertheless, with ultrashort laser pulses it may be possible to create transient inversions for amplifying ultrashort VUV pulses. For example, if the (2,0) state is populated at the same time as the (1,1) state of I_2^{2+}, there will be no inversion at first, as more population ends up in the lower (1,1) state. However, as wavepackets are formed on the two states, they will start to dissociate, although the speed of dissociation will be different for the two states. Thus, at a later time, the population on the two curves will no longer overlap in space and an inversion on the (2,0) to (1,1) transition can occur at a particular R, as shown in Fig. 1.18.

The I_2^{2+} state provides another interesting example of how to form an inversion. As was discussed earlier, the (2,0) state actually supports a bound

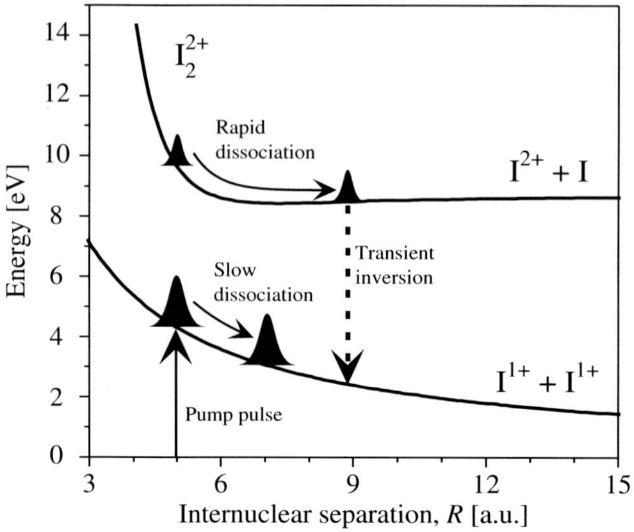

Fig. 1.18. Example of transient inversions that form during short-pulse ionization of molecules

potential well. If bound vibrational states are populated, their only decay path is radiative, which will occur on the 100 ps time scale. In the meantime, any population in the (1,1) state will have dissociated, again leaving an inversion, although this inversion would persist for hundreds of picoseconds. Again, see Fig. 1.18. This is very similar to the rare gas halide excimers, which are vibrationally bound in the excited state and unbound in the ground state, allowing inversions to form [45].

1.5 Conclusions

Molecules in strong laser fields display a rich variety of behaviors, including a number of different mechanisms for creating vibrational and electronic excitation. Often these types of excitation are coupled and can provide insight into both the electronic structure of the molecule and how its response to strong laser fields depends on internuclear separation and spatial orientation. Understanding these excitation mechanisms is important for determining their effect on high-harmonic generation and quantum tomography. Moreover, these interactions may lead to some novel forms of coherent control and the creation of population inversions in the VUV spectral region.

Acknowledgements

The authors acknowledge continued support from the NSF for this work, most recently from Grant No. PHYS-0653029.

References

1. P.B. Corkum, Phys. Rev. Lett. **71**, 1994 (1993)
2. B. Walker, B. Sheehy, L.F. DiMauro, P. Agostini, K.J. Schafer, K.C. Kulander, Phys. Rev. Lett. **73**, 1227 (1994)
3. B. Walker, B. Sheehy, K.C. Kulander, L.F. DiMauro, Phys. Rev. Lett. **77**, 5031 (1996)
4. A. LHuillier, P. Balcou, Phys. Rev. Lett. **70**, 774 (1993);**** J.J. Macklin, J.D. Kmetec, C.L. Gordon, Phys. Rev. Lett. **70**, 766 (1993)
5. I.V. Litvinyuk, K.F. Lee, P.W. Dooley, D.M. Rayner, D.M. Villeneuve, P.B. Corkum, Phys. Rev. Lett. **90**, 233003 (2003)
6. M. Ivanov, T. Seideman, P. Corkum, F. Ilkov, P. Dietrich, Phys. Rev. A **54**, 1541 (1996)
7. K. Boyer, T.S. Luk, J.C. Solem, C.K. Rhodes, Phys. Rev. A **39**, 1186 (1989)
8. R.J. Verver, D.R. Matusek, J.S. Wright, G.N. Gibson, R. Bhardwaj, S. Aseyev, D.M. Villeneuve, P.B. Corkum, M.Y. Ivanov, J. Phys. Chem. A **105**, 2435 (2001)
9. J.P. Nibarger, M. Li, S. Menon, G.N. Gibson, Phys. Rev. Lett. **83**, 4975 (1999)
10. M. Lein, P.P. Corso, J.P. Marangos, P.L. Knight, Phys. Rev. A **67**, 023819 (2003)
11. J.H. Posthumus, K. Colding, L.J. Frasinski, M.R. Thompson, Laser Phys. **7**, 813 (1997)
12. S.S. Viftrup, V. Kumarappan, S. Trippel, H. Stapelfeldt, E. Hamilton, T. Seideman, Phys. Rev. Lett. **99**, 143602 (2007)
13. J.J. Larsen, K. Hald, N. Bjerre, H. Stapelfeldt, T. Seideman, Phys. Rev. Lett. **85**, 2470 (2000)
14. G.N. Gibson, R.R. Freeman, T.J. McIlrath, Phys. Rev. Lett. **67**, 1230 (1991)
15. R.N. Coffee, G.N. Gibson, Phys. Rev. A **69**, 053407 (2004)
16. L. Fang, G.N. Gibson, Phys. Rev. Lett. **100**, 103003 (2008)
17. L. Fang, G.N. Gibson, Phys. Rev. A **75**, 063410 (2007)
18. T.E. Sharp, Atom. Data **2**, 119 (1971)
19. D. Villarejo, J. Chem. Phys. **49**, 2523 (1968)
20. A. Zavriyev, P.H. Bucksbaum, H.G. Muller, D.W. Schumacher, Phys. Rev. A **42**, 5500 (1990); A. Zavriyev, P.H. Bucksbaum, J. Squier, F. Saline, Phys. Rev. Lett. **70**, 1077 (1993)
21. T. Ergler, A. Rudenko, B. Feuerstein, K. Zrost, C.D. Schrőter, R. Moshammer, J. Ullrich, Phys. Rev. Lett. **97**, 193001 (2006); I.A. Bocharova, H. Mashiko, M. Magrakvelidze, D. Ray, P. Ranitovic, C.L. Cocke, I.V. Litvinyuk, Phys. Rev. A **77**, 053407 (2008)
22. J.P. Nibarger, S.V. Menon, G.N. Gibson, Phys. Rev. A **63**, 053406 (2001)
23. S.V. Menon, J.P. Nibarger, G.N. Gibson, J. Phys. B **35**, 2961 (2002)
24. A. Lofthus, P.H. Krupenie, J. Phys. Chem. Ref. Data **6**, 113 (1977)
25. W.A. de Jong, L. Visscher, W.C. Nieuwpoort, J. Chem. Phys. **107**, 9046 (1997); J. Che, J.L. Krause, M. Messina, K.R. Wilson, Y. Ya, J. Phys. Chem. **99**, 14949 (1995); M.C.R. Cockett, R.J. Donovan, K.P. Lawley, J. Chem. Phys. **105**, 3347 (1996)
26. G. Gibson, T.S. Luk, C.K. Rhodes, Phys. Rev. A **41**, 5049 (1990)
27. T. Ergler, B. Feuerstein, A. Rudenko, K. Zrost, C.D. Schroter, R. Moshammer, J. Ullrich, Phys. Rev. Lett. **97** 103004 (2006)

28. E. Goll, G. Wunner, A. Saenz, Phys. Rev. Lett. **97** 103003 (2006); A. Saenz, J. Phys. B **33**, 4365 (2000)
29. L. Fang and G. N. Gibson, Phys. Rev. A, Rapid Comm. **78**, 051402(R) (2008)
30. L.J. Frasinski, J. Plumridge, J.H. Posthumus, K. Codling, P.F. Taday, E.J. Divall, A.J. Langley, Phys. Rev. Lett. **86**, 2541 (2001)
31. D.T. Strickland, Y. Beaudoin, P. Dietrich, P.B. Corkum, Phys. Rev. Lett. **68**, 2755 (1992)
32. D. R. Lide *Handbook of Chemistry and Physics*, 73rd edn. (CRC Press, Boca Raton, 1992), pp. 10-211–10-213.
33. G.N. Gibson, M. Li, C. Guo, J.P. Nibarger, Phys. Rev. A. **58**, 4723 (1998)
34. C. Guo, M. Li, G.N. Gibson, Phys. Rev. Lett. **82**, 2492 (1999)
35. R.S. Mulliken, J. Chem. Phys. **7**, 20 (1939)
36. G.N. Gibson, Phys. Rev. A **67**, 043401 (2003)
37. G.N. Gibson, Phys. Rev. Lett. **89**, 263001 (2002)
38. W.D. Oliver, Y. Yu, J.C. Lee, K.K. Berggren, L.S. Levitov, T.P. Orlando, Science **310**, 1653 (2005)
39. G.N. Gibson, R.N. Coffee, L. Fang, Phys. Rev. A **73**, 023418 (2006)
40. G.N. Gibson, L. Fang, B. Moser, Phys. Rev. A **74**, 041401(R) (2006)
41. A.S. Alnaser, M. Zamkov, X.M. Tong, C.M. Maharjan, P. Ranitovic, C.L. Cocke, I.V. Litvinyuk, Phys. Rev. A **72**, 041402 (2005)
42. G. Gibson, T.S. Luk, A. McPherson, K. Boyer, C.K. Rhodes, Phys. Rev. A **40**, 2378 (1989)
43. S. Voss, A.S. Alnaser, X.-M. Tong, C. Maharjan, P. Ranitovic, B. Ulrich, B. Shan, Z. Chang, C.D. Lin, C.L. Cocke, J. Phys. B **37**, 4239 (2004)
44. J. Itatani, J. Levesque, D. Zeidler, H. Niikura, H. Pépin, J.C. Kieffer, P.B. Corkum, D.M. Villeneuve, Nature **432**, 867 (2004)
45. C.A. Brau, in *Excimer Lasers*, ed. by C.K. Rhodes. Topics in Applied Physics, vol 30 (Springer, Berlin, 1984)

2

Coherent Lattice Oscillations in Solids and Their Optical Control Part I. Fundamentals and Optical Detection Techniques

Kunie Ishioka and Oleg V. Misochko

Summary. Coherent optical phonons are the lattice atoms vibrating in phase with each other over a macroscopic spatial region. With sub-10 fs laser pulses, one can impulsively excite the coherent phonons of a frequency up to 50 THz, and detect them optically as a periodic modulation of electric susceptibility. The generation and relaxation processes depend critically on the coupling of the phonon mode to photoexcited electrons. Real-time observation of coherent phonons can thus offer crucial insight into the dynamic nature of the coupling, especially in extremely nonequilibrium conditions under intense photoexcitation.

2.1 Introduction

Illumination of light on solids leads to electron–hole pair excitation if the photon energy exceeds the bandgap. The photoexcited carriers cool down in femto- to picoseconds, first by distributing the excess energy among themselves and then by transferring the energy to the lattice sub-system. The latter process has been extensively studied in semiconductors by time-resolved Raman spectroscopy. Because Raman scattering requires a nearly monochromatic incident light, the time-resolution has been limited to the picosecond range.

With development of ultrashort pulsed lasers, coherently generated lattice dynamics was found, first as the periodic modulation in the transient grating signal from perylene in 1985 by De Silvestri and coworkers [1]. Shortly later, similar modulation was observed in the reflectivity of Bi and Sb [2] and of GaAs [3], as well as in the transmissivity of YBCO [4] by different groups. Since then, the coherent optical phonon spectroscopy has been a simple and powerful tool to probe femtosecond lattice dynamics in a wide range of solid

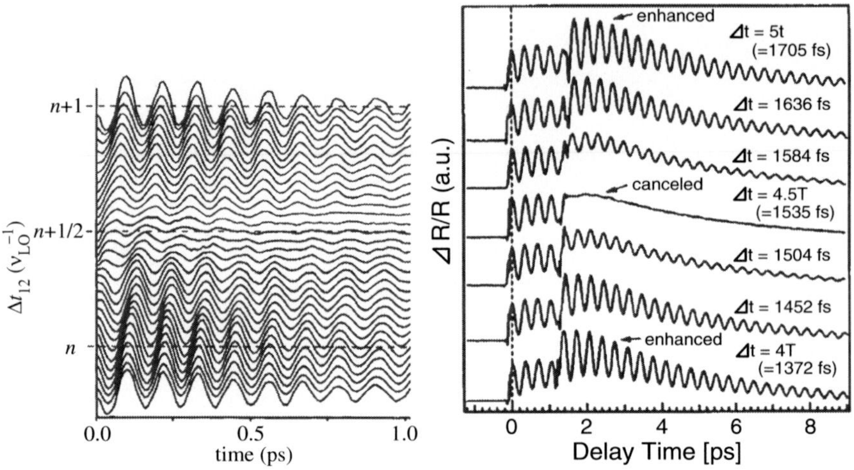

Fig. 2.1. Double-pump and probe reflectivity signal of coherent LO phonons of GaAs (*left*) and coherent A_{1g} phonons of Bi (*right*). The horizontal axis gives the time delay of the probe pulse with respect to the first pump pulse in both panels. Δt_{12} (Δt) in the *left* (*right*) panel is the time delay between the two pump pulses in units of the LO phonon period ν_{LO}^{-1}=114 fs (A_{1g} phonon period $T = 341$ fs). From [5] and [6]

materials. The coherent optical phonons are generated only during the illumination of femtosecond optical pulses, in contrast to hot incoherent phonons created during hot carrier cooling for picoseconds. They are coherent in the sense that two time-separated optical pulses can enhance or destroy the oscillation depending on the separation between the pulses (Fig. 2.1). In contrast to coherent acoustic phonons propagating as ballistic strain pulses in crystals, the coherent phonons of optical branches are nonpropagating and delocalized, as light couples directly with phonons only at the Γ point of the Brillouin zone ($k = 0$) where the group velocity is nearly zero.

This and the next chapters give an overview of the recent experimental researches on coherent optical phonons. The generation mechanism and the optical detection techniques are described in Sects. 2.2 and 2.3 of this chapter, respectively, followed by the findings of the optical experiments in a variety of solid materials in Sects. 2.4–2.7. The next chapter presents the non-optical detection techniques and their findings in Sect. 3.1, together with the achievements of the optical control experiments in Sect. 3.2. As we would like to stress recent developments in limited pages, we omit many classical works on which the readers can refer to the excellent reviews on ferroelectrics [7,8], GaAs and its heterostructures [9,10], π-conjugated chains [11], dihalogen-doped rare gas crystals [12] and quasi-1D halogen-bridged metal complexes [13], and surface phonons [14,15].

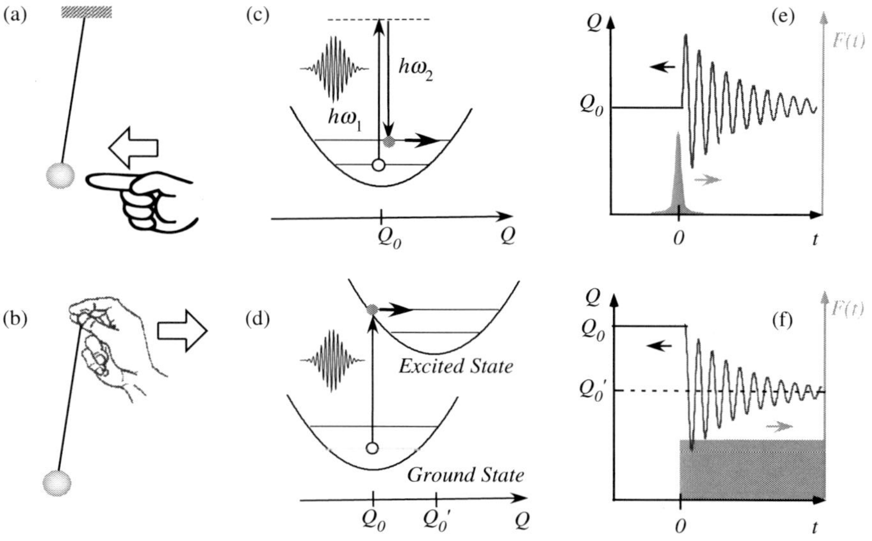

Fig. 2.2. Two generation models of coherent optical phonons. (**a**), (**c**), (**e**) impulsive stimulated Raman scattering (ISRS). (**b**), (**d**), (**f**) displacive excitation of coherent phonons (DECP). Graphs (**e**) and (**f**) display the time evolution of the driving force (*grey areas*) and that of the displacement (*solid curves*) for ISRS and DECP, respectively

2.2 Generation of Coherent Phonons

The classical equation of motion[1] describing the coherent phonons for a small nuclear displacement Q is that of a driven harmonic oscillator [9, 10, 15]

$$\mu\left[\frac{\partial^2 Q(t)}{\partial t} + 2\gamma\frac{\partial Q(t)}{\partial t} + \omega_0^2 Q(t)\right] = F(t), \tag{2.1}$$

with the reduced lattice mass μ, the damping γ and the frequency ω. The generation mechanism of the coherent phonons can be classified into two types, "impulsive" and "displacive," according to the temporal-profile and the origin of the force $F(t)$. The impulsive excitation is realized with a δ-function-like driving force created via stimulated Raman scattering (Fig. 2.2 c, e). Displacive excitation is achieved by a step-function-like force via a potential shift in the excited state (Fig. 2.2 d, f).

2.2.1 Impulsive Stimulated Raman Scattering

Up to now, only Raman active modes at the Γ point of the Brillouin zone have been observed as coherent phonons in bulk crystals.[2] The selection rule can be

[1] Quantum description of coherent vibrations is given in [7, 15, 16].

[2] Exceptions include coherent IR-active (but Raman-inactive) phonons observed as a THz emission from Te [17] and CdTe [18].

partly explained in terms of impulsive stimulated Raman scattering (ISRS) [7, 15, 19] in the generation process, illustrated in Fig. 2.2c for a nonresonant case. A broadband optical pulse can offer multiple combinations of two photons whose energy difference matches the vibrational energy ($h\omega_1 - h\omega_2 = h\omega_0$), and give an impulsive driving force to initiate coherent nuclear motion on the ground electronic state. This situation is an analogue of giving a kick to a weight of a pendulum and changing its kinetic energy (Fig. 2.2a). The driving force is defined by the Raman polarizability [9, 15]

$$F(t) = \frac{1}{2}\left(\frac{\partial \alpha}{\partial Q}\right)_{kl} E_k E_l. \tag{2.2}$$

The key requirements for ISRS excitation are the existence of Raman active phonons in the crystal, and the pulse duration shorter than the phonon period ω_0^{-1} [19]. The resulting nuclear oscillation follows a sine function of time (i.e., minimum amplitude at t=0), as shown in Fig. 2.2e. ISRS occurs both under nonresonant and resonant excitations. As the Raman scattering cross section is enhanced under resonant excitation, so is the amplitude of the ISRS-generated coherent phonons.

Experimental verification of the ISRS generation can be primarily given by the pump polarization dependence. The coherent phonons driven by ISRS (second order process) should follow the symmetry of the Raman tensor, while those mediated by photoexcited carriers should obey the polarization dependence of the optical absorption (first order process). It is possible, however, that both ISRS and carrier-mediated generations contribute to the generation of a single phonon mode. The polarization dependence is then described by the sum of the first- and second-order processes [20–22], as shown in Fig. 2.3.

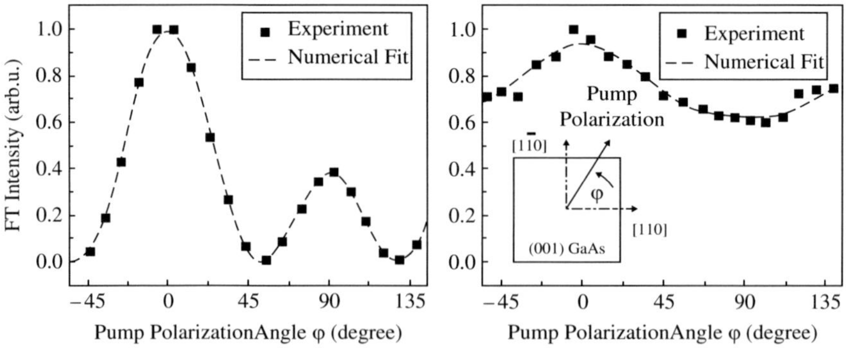

Fig. 2.3. The Fourier-transformed (FT) intensity of coherent phonons as a function of the pump polarization angle φ for a GaAs/$Al_{0.36}Ga_{0.64}As$ MQW. The excitation wavelength is slightly above the $n=1$ exciton resonance (*left*) and slightly above the $n=2$ subband energy (*right*). φ-dependent component is attributed to ISRS, while the φ-independent component is to TDFS and forbidden Raman scattering. From [20]

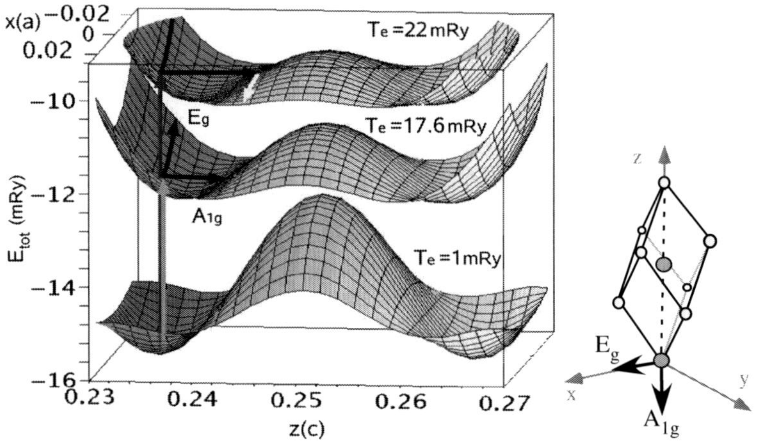

Fig. 2.4. *Left*: potential energy surfaces calculated for the ground state ($T_e = 1$ mRy) and for excited electronic states ($T_e = 17.6$ and 22 mRy) of Bi. *Vertical arrows* show transitions of the atoms from the ground state to an excited energy surface. From [23]. *Right*: crystalline structure of Bi and Sb crystals. The directions of the A_{1g} and E_g modes are indicated in both panels

2.2.2 Photocarrier-Mediated Excitation of Coherent Phonons

Early pump-probe experiments on semimetals Bi and Sb by Cheng and coworkers detected only the fully symmetric (A_{1g}) mode but not the other Raman active (E_g) mode [2,24]. The apparent absence of the E_g mode[3] led the authors to propose an alternative generation mechanism: displacive excitation of coherent phonons (DECP) [27]. In DECP, a sudden shift of the vibrational potential kick-starts the coherent oscillation on the electronic excited state (Fig. 2.2d). The shift in the equilibrium coordinate $(Q_0'(t) - Q_0)$ is assumed to be a linear function of the photoexcited electron density $n(t)$. This situation is analogous to quickly translating the suspension point of a pendulum and thereby changing the potential energy (Fig. 2.2b). The requirements for DECP generation are such electron–phonon coupling that shifts the potential energy surface by electronic excitation, and the pulse duration shorter than the phonon period. The resulting nuclear oscillation is a cosine function of time (i.e., maximum amplitude at $t = 0$), as shown in Fig. 2.2f.

The DECP model successfully explained the observed initial phase of the fully symmetric phonons in a number of opaque crystals [24]. The absence of the E_g mode was attributed to an exclusive coupling between the electrons photoexcited near the Γ point and the fully symmetric phonons. A recent density functional theory (DFT) calculation [23] demonstrated this exclusive coupling as the potential energy surface (Fig. 2.4). The minimum of the potential surface of the excited state shifted significantly along the trigonal (z) axis,

[3] Coherent E_g phonon was later observed experimentally in Bi [6,25] and Sb [26].

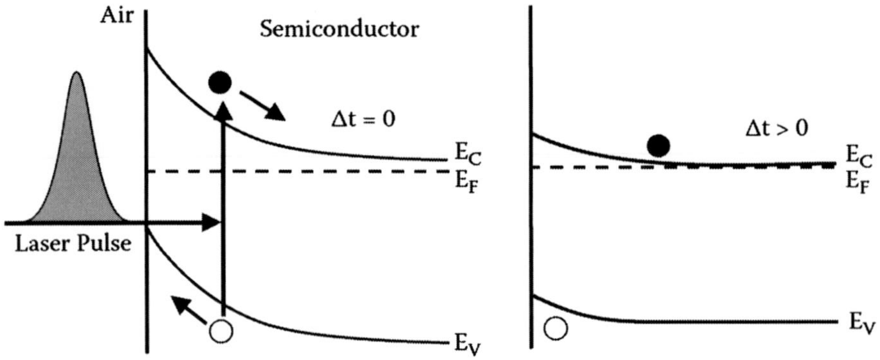

Fig. 2.5. Schematic sketch of transient depletion field screening (TDFS) in an n-doped semiconductor with midgap Fermi-level pinning at the surface. Excitation of free electron–hole pairs leads to rapid drift currents on the time scale of the exciting laser pulse (*left*). After the excitation, electrons and holes are spatially separated and the space charge field is screened (*right*). From [10]

that is, the direction of the A_{1g} mode, because it is the direction of the Peierls distortion. In contrast, the minimum shifts very little in the direction of the E_g mode, which is perpendicular to the direction of the Peierls distortion.

As the lattice interacts with light only through electrons, both DECP and ISRS should rely on the electron–phonon coupling in the material. Distinction between the two models lies solely in the nature of the electronic transition. In this context, Merlin and coworkers proposed DECP to be a resonant case of ISRS with the excited state having an infinitely long lifetime [26, 28]. This original "resonant ISRS" model failed to explain different initial phases for different coherent phonon modes in the same crystal [21, 25]. Recently, the model was modified to include finite electronic lifetime [29] to have more flexibility to reproduce the experimental observations.

In polar semiconductors, carrier-mediated generation occurs in the form of transient depletion field screening (TDFS) depicted in Fig. 2.5 [30]. The driving force in (2.1) can be expressed by the sum of the Raman term and the nonlinear longitudinal polarization [10]:

$$F_j(t) = \frac{1}{2}\left(\frac{\partial\alpha}{\partial Q}\right)_{kl} E_k E_l - \frac{e^*}{\epsilon_\infty \epsilon_0} P_j^{NL}. \tag{2.3}$$

The nonlinear polarization can be divided into several different contributions:

$$P_j^{\mathrm{NL}} = \chi_{jkl}^{(2)} E_k E_l + \chi_{jklm}^{(3)} E_k E_l E_m + \int_{-\infty}^{t} \mathrm{d}t' J_j(t')$$
$$+ Ne \int_{-\infty}^{t} \mathrm{d}t' \int_{-\infty}^{\infty} \mathrm{d}x_j \langle \Psi(x_j, t') | x_j | \Psi(x_j, t') \rangle. \tag{2.4}$$

The third term describes the polarization set up by ultrafast drift-diffusion currents, which can excite coherent phonons via TDFS (or via the buildup of electric Dember fields [9, 10]). The first two terms represent the second- and the third-order nonlinear susceptibilities, respectively [31]. The fourth term describes the polarization associated with coherent electronic wavefunctions, which becomes important in semiconductor heterostructures.

For surface coherent phonons of ferromagnetic metals, a spin-driven generation mechanism was proposed, as will be described in Sect. 2.6.

2.3 Optical Detection of Coherent Phonons

Optical detection offers the most conventional technique to time-resolve the coherent phonons. It includes four-wave mixing [8], transient reflectivity [9,10] and transmission [7] measurements, as well as second harmonic generation (SHG) [15, 32]. Coherent nuclear displacement Q induces a change in the optical properties (e.g., reflectivity R) of the crystal through the refractive index n and the susceptibility χ,

$$\Delta R = \frac{\partial R}{\partial n}\Delta n \approx \frac{\partial R}{\partial \chi}\frac{\partial \chi}{\partial Q}Q(t), \tag{2.5}$$

which makes ΔR to be a quantitative, though not absolute, measure for Q. Because $\partial\chi/\partial Q$ is a first-order Raman tensor, only Raman active modes with nonzero $\partial\chi/\partial Q$ can be detected by linear optical detection [9, 10].

Figure 2.6 illustrates an experimental setup for pump-probe reflectivity measurements. In the standard reflectivity measurements, the reflected probe light is detected without polarization analysis. In this scheme, coherent phonons of any symmetry can in principle be detected, given the polarizations of the pump and probe beams appropriately chosen. By contrast, the "anisotropic" scheme, often referred to as "electro-optic (EO) sampling" [33], detects only the anisotropic component of the reflectivity arising from non-fully symmetric phonons. It employs probe beam polarized at 45° with respect to the optical plane, and detection of the difference of the s- and p-polarized components of the reflected probe beam ($\Delta R_p - \Delta R_s$) with a pair of matched photodiodes, as illustrated in Fig. 2.6. Another technique illustrated is the fast scan, in which the signal is stored and averaged in a digital oscilloscope while the pump-probe delay is continuously scanned. This enables an averaging of the signal over 1,000 times in a minute, and thus improves the signal-to-noise ratio drastically. With these techniques combined, one can detect a reflectivity change as small as $\Delta R/R \sim 10^{-6}$.

By using a nonlinear optical process such as SHG, one can probe surface phonons and adsorbate-related vibrations exclusively [14, 15, 32, 34]. Time-resolved SHG (TRSHG) detects the second harmonic (SH) of the probe beam as a function of time delay between pump and probe. The SH electric field is driven by the nonlinear polarization $P_i(2\omega)$ at the surface, which

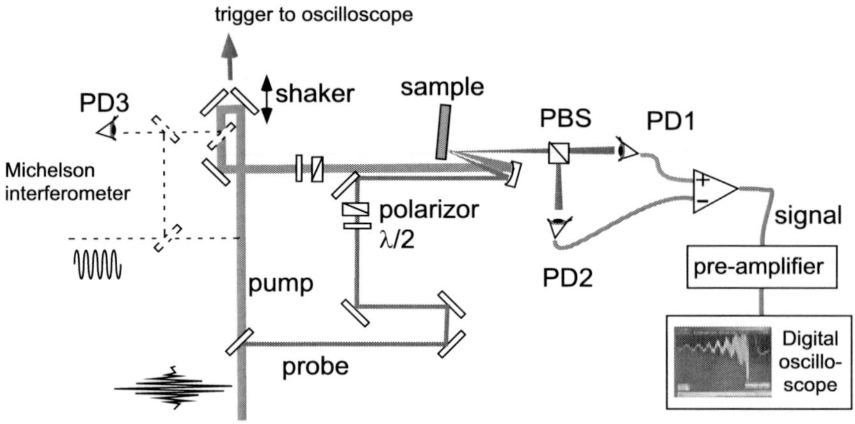

Fig. 2.6. Schematic illustration of the experimental setup for pump-probe anisotropic reflectivity measurements with fast scan method. PBS denotes polarizing beam splitter, PD1 and PD2, a pair of matched photodiodes to detect *p*- and *s*-polarized components of the reflected probe beam, PD3 another photodiode to detect the interference pattern of He–Ne laser in a Michelson interferometer to calibrate the scanning of the pump path length

can be expanded to the first order of the nuclear displacement $Q_{n,q}$ for the n-th phonon mode with wave vector q [14]:

$$P_i(2\omega) = \left[\chi^{(2)}_{ijk}(2\omega) + \sum_{n,q} \left(\partial\chi^{(2)}_{ijk}(2\omega)/\partial Q_{n,q}\right)_0 Q_{n,q}\right] \times E_j(\omega)E_k(\omega). \quad (2.6)$$

The SH intensity is proportional to $|P|^2$. Experimentally, the oscillatory part of the total SH is so small that one can ignore its second-order term. If coherent surface phonons are created by ISRS, the whole process including excitation and detection is the coherent time-domain analogue of stimulated hyper Raman scattering ($\chi^{(4)}$ process) [14]. The cross section of the SHG process is then proportional to the product of a Raman tensor in the pump transition and a hyper-Raman tensor $\partial\chi^{(2)}_{ijk}/\partial Q_n$ in the probe transition.

2.4 Electron–Phonon Coupling in Group V Semimetals

Semimetals bismuth (Bi) and antimony (Sb) have been model systems for coherent phonon studies. They both have an A7 crystalline structure and sustain two Raman active optical phonon modes of A_{1g} and E_g symmetries (Fig. 2.4). Their pump-induced reflectivity change, shown in Fig. 2.7, consists of oscillatory ($\Delta R_{\rm osc}$) and non-oscillatory ($\Delta R_{\rm nonosc}$) components. $\Delta R_{\rm osc}$ is dominated by the coherent nuclear motion of the A_{1g} and E_g symmetries, while $\Delta R_{\rm nonosc}$ is attributed to the modification in the electronic and the lattice temperatures.

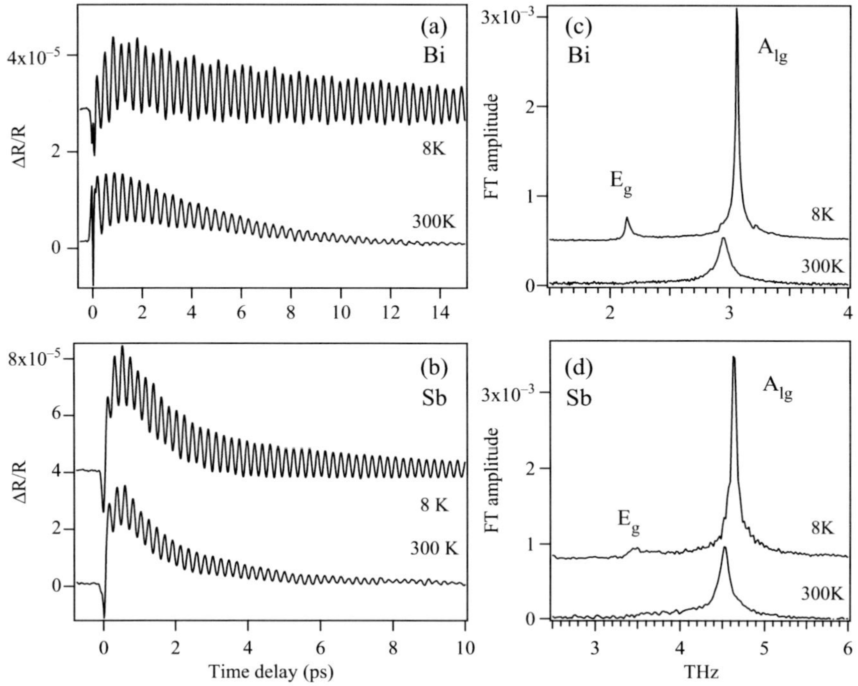

Fig. 2.7. (**a**, **b**): transient reflectivity changes $\Delta R/R$ of Bi and Sb single crystals at different temperature. (**c**, **d**): FT spectra of the oscillatory component $\Delta R_{osc}/R$. The excitation pulse energy is $9\,\mu J/cm^2$ for Bi and $56\,\mu J/cm^2$ for Sb. The traces are offset for clarity. Adapted from [25] and [21]

The coherent oscillation of the A_{1g} mode was a cosine function of time (Fig. 2.8) [21, 24–26]. Recent X-ray measurements demonstrated a clear shift in the equilibrium position at photoexcitation [35], as we will see in the next chapter. These results confirmed the displacive generation of coherent A_{1g} phonons, as discussed in Sect. 2.2.2. In contrast, the coherent oscillation of the E_g mode was a sine function of time, and its amplitude exhibited a $\cos 2\varphi$ dependence on the pump polarization angle φ. Both features indicated the ISRS generation of the coherent E_g phonons [21, 25].

Under moderate ($\mu J/cm^2$) photoexcitation, where the photoexcited carrier density is comparable or less than the intrinsic density, time evolution of coherent A_{1g} and E_g phonons is respectively described by a damped harmonic oscillation

$$Q(t) \propto \frac{\Delta R_{\text{osc}}}{R} = A_0 \exp(-\Gamma_0 t) \sin(\omega_0 t + \delta), \tag{2.7}$$

with the dephasing rate Γ_0 and the frequency ω_0 being independent of the photoexcitation density [21, 25]. Under intense (mJ/cm^2) photoexcitation, in contrast, both the A_{1g} and E_g phonons dephased faster, and their

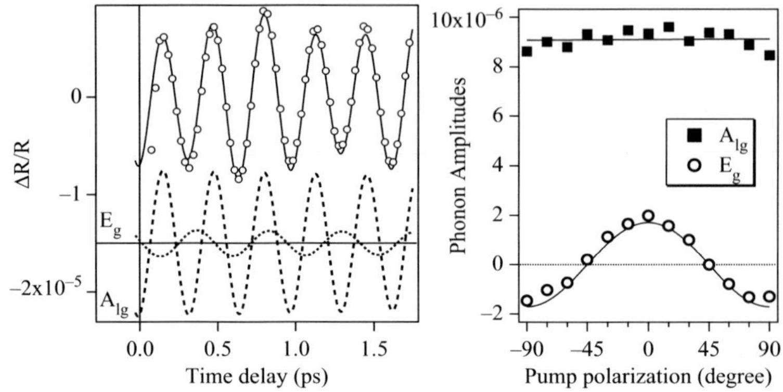

Fig. 2.8. *Left*: oscillatory part of the reflectivity change of Bi (0001) surface at 8K (*open circles*). Fit to the double damped harmonic function (*solid curve*) shows that the A_{1g} and E_g components (*broken* and *dotted curves*) are a sine and a cosine functions of time, respectively. *Right*: pump polarization dependence of the amplitudes of coherent A_{1g} and E_g phonons of Bi (0001). Adapted from [25]

frequencies redshifted, with increasing excitation density [36–38]. In this strong-excitation regime, the coherent phonons are better described as *chirped* damped oscillations

$$\frac{\Delta R_{\mathrm{osc}}}{R} = A_0 \exp(-\Gamma_0 t) \sin((\omega_0 + \alpha t)t + \delta), \tag{2.8}$$

with the frequency varying at a positive rate α with time delay (Fig. 2.9). The origin of the redshift was first attributed to the lattice anharmonicity by Hase and coworkers [36], but later double-pulse experiments [39] combined with DFT calculations [40] by Murray and coworkers attributed it to the electronic softening. In fact, the redshift in the frequency and the increase in the dephasing rate can be understood in terms of a Fano interference between two paths: one directly from the discrete (phonon) state and another mediated by a (electronic) continuum [38, 41]. The interference leads to an asymmetric line shape in the frequency domain given by

$$f(\epsilon) = \frac{(\epsilon + q)^2}{1 + \epsilon^2}, \tag{2.9}$$

with q being the asymmetry parameter and ϵ the dimensionless energy. The parameter q is a function of the phonon self-energy, whose real and imaginary parts correspond to the bare phonon frequency and the broadening, respectively. Its temperature- and pump power-dependences indicated that the continuum includes both the lattice and electronic degrees of freedom [38]. The contribution of the lattice anharmonicity was also confirmed by the observation of the two-phonon combination modes (e.g., $A_{1g} - E_g$ and $A_{1g} + E_g$) in the experiment [42] and theory [23], though the electronic softening undoubtedly plays a substantial role.

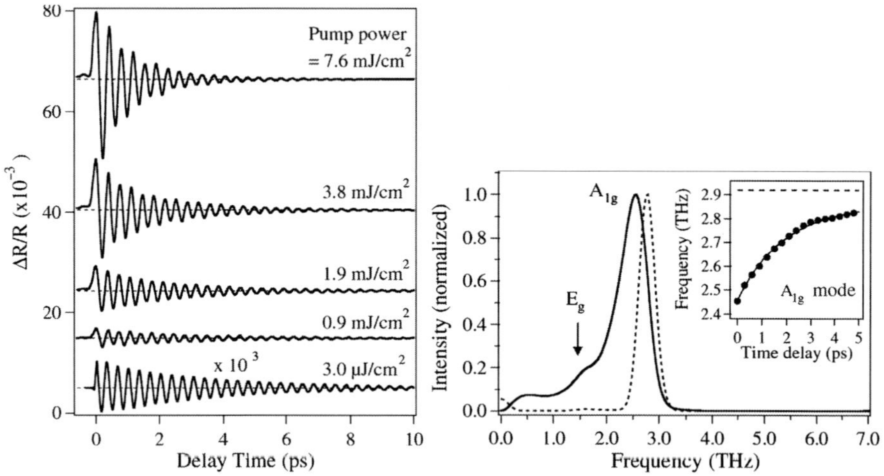

Fig. 2.9. *Left*: transient reflectivity change of Bi at various pump densities. *Right*: discrete wavelet transformation spectra obtained for time delay of 0.3 ps (*solid line*) and 3.0 ps (*dotted line*). Inset in right panel shows the A_{1g} frequency as a function of the time delay. The *dashed line* in inset represents the equilibrium frequency. From [36]

Under even more intense photoexcitation ($\sim$10 mJ/ cm^2), the coherent A_{1g} and E_g phonons of Bi and Sb exhibit a collapse-revival in their amplitudes (Fig. 2.10) [42,43]. This phenomenon has a clear threshold in the pump density, which is common for the two phonon modes but depends on temperature and the crystal (Bi or Sb). At first glance, the amplitude collapse-revival appears to be analogous to the fractional revival in nuclear wavepackets in molecules [44, 45]. However, the pump power dependence may be an indication of a polarization, not quantum, beating between different spatial components of the coherent response within the laser spot [46].

2.5 Coherent Phonons in Group IV Crystals and Graphitic Materials

2.5.1 Coherent Phonons in Tetrahedrally Bonded Crystals

Silicon is a model for the fundamental electronic and mechanical properties of Group IV crystals and the basic material for electronic device technology. Coherent optical phonons in Si revealed the ultrafast formation of "renormalized" quasiparticles in time-frequency space [47]. The anisotropic transient reflectivity of n-doped Si(001) featured the coherent optical phonon oscillation with a frequency of 15.3 THz, when the [110] crystalline axis was parallel to the pump polarization (Fig. 2.11). Rotation of the sample by 45° led to disappearance of the coherent oscillation, which confirmed the ISRS generation,

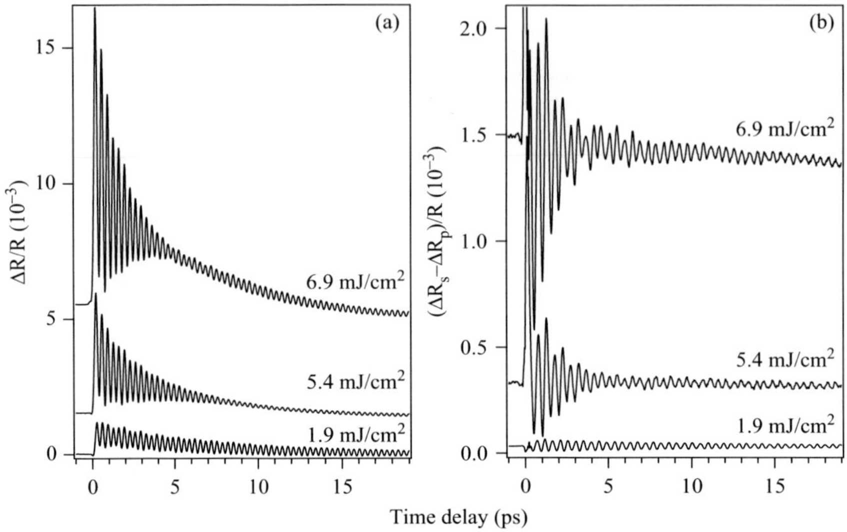

Fig. 2.10. Transient reflectivity change $\Delta R/R$ (**a**) and its anisotropic component $(\Delta R_p - \Delta R_s)/R$ (**b**) from Bi (0001) at 8 K for different pump densities. Pump polarization is chosen to show only the coherent A_{1g} or E_g phonons, respectively

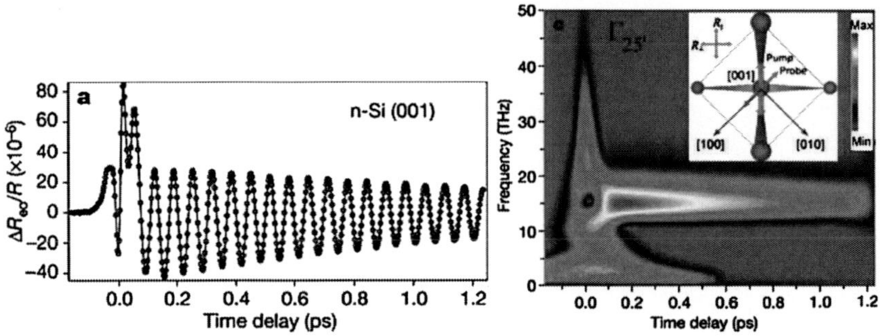

Fig. 2.11. Transient anisotropic reflectivity change for Si(001) in the $\Gamma_{25'}$ geometry (*left*) and its continuous wavelet transform (*right*). Inset in the right panel defines the polarization of the pump beam relative to the crystalline axes. From [47]

like in Ge [48]. Time-dependent spectral amplitude (chronogram) in Fig. 2.11 revealed the fast broadband response near $t = 0$ due to the coherent electronic coupling of the pump and probe fields via the nonlinear susceptibility. The optical phonon was seen at 15.3 THz for all the time delay $t > 0$. The most intriguing aspect of the chronogram was the anti-resonance at 15.3 THz slightly after $t = 0$, suggesting the coherent excitation of coupled phonon–carrier systems via electronic Raman process possibly near the Γ point and along the Λ direction. The data in Fig. 2.11 revealed interference effects leading to the coherent phonon generation and subsequent “dressing” by electron–hole

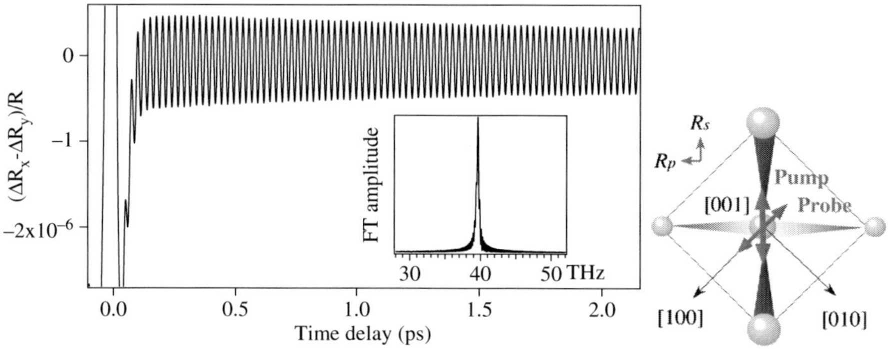

Fig. 2.12. *Left*: transient anisotropic reflectivity change of the (001) surface of single crystal type IIa diamond. Inset shows the FT spectrum of the oscillation, demonstrating a narrow peak of the optical phonon at 40 THz. *Right*: pump and probe polarizations to detect the optical phonon. Adapted from [50]

pairs. They clearly demonstrated the possibility of observing the quantum mechanical manifestations of carrier–phonon interactions in the real time, which until now could only be deduced from transport measurements and spectral lineshape analysis.

Isotope superlattices of nonpolar semiconductors gave an insight on how the coherent optical phonon wavepackets are created [49]. High-order coherent confined optical phonons were observed in $^{70}Ge/^{74}Ge$ isotope superlattices. Comparison with the calculated spectrum based on a planar force-constant model and a bond polarizability approach indicated that the coherent phonon amplitudes are determined solely by the degree of the atomic displacement, and that only the Raman active odd-number-order modes are observable.

Diamond has the widest bandgap among the group IV crystals. The anisotropic reflectivity change of diamond was modified at a period of 25 fs (frequency of 40 THz) when the pump polarization was parallel to the [110] direction (Fig. 2.12) [50]. The oscillatory signal essentially vanished after rotating the sample by 45°, as in Si [47] and Ge [48], which confirms the Raman generation. The dephasing rate of the optical phonon was obtained to be $\Gamma_0 = 0.145\,\mathrm{ps}^{-1}$, giving a very large Q factor of ∼300. The amplitude of the coherent oscillation showed a linear pump power dependence. It clearly indicated the off-resonance, field-driven nature of the coherent phonon generation with 3.1 eV photons, whose energy was much smaller than the indirect (5.48 eV) or direct (7.3 eV) bandgaps.

2.5.2 Ultrafast Electron–Phonon Decoupling in Graphite

Graphite possesses highly anisotropic layered crystal structure, which translates to a quasi-2D electronic structure with electronic bands dispersing linearly near E_F and forming point-like Fermi surfaces. Visible light induces

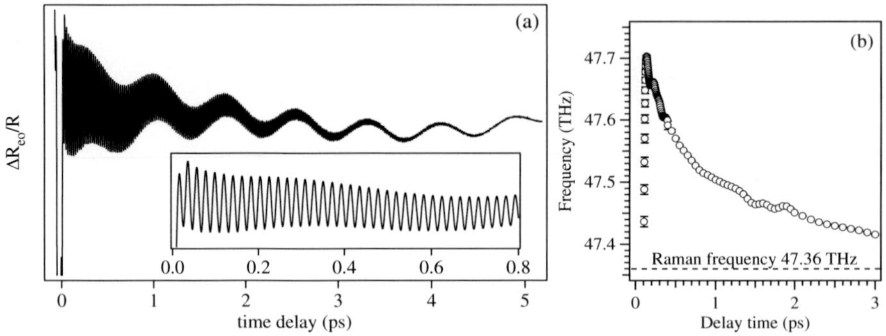

Fig. 2.13. (**a**) Anisotropic reflectivity change of graphite measured with sub-10 fs pulses at 3.1 eV. (**b**) Time evolution of the E_{2g2} phonon frequency obtained from time-windowed FT. From [51]

vertical transitions from the valence (π) to the conduction (π^*) bands near the K point of the Brillouin zone. Anisotropic reflectivity change of graphite (Fig. 2.13a) was modulated at two disparate periods of 21 and 770 fs, assigned to the in-plane carbon stretching (Raman G peak) [51] and the interlayer shear mode [52], respectively. The amplitudes of both phonons exhibited a $\cos 2\varphi$ dependence on the pump polarization angle φ, confirming their ISRS generation.

The coherent in-plane (E_{2g2}) phonon of graphite has a frequency of 47 THz. A time-windowed FT analysis (Fig. 2.13b) revealed that the phonon frequency blue-shifted promptly at the photoexcitation [51], contrary to the coherent phonon response in Bi, Sb (Sect. 2.4), Zn or Zr (Sect. 2.6). The recovery of the frequency to its near-equilibrium value (47.36 THz) took several picoseconds. The time scale, corresponding to those of carrier thermalization and carrier-lattice equilibration, suggests that the frequency shift is dominated by the excited electron dynamics. A DFT calculation demonstrated that non-adiabaticity plays an essential role in the phonon stiffening [51]. For unexcited graphite, the E_{2g2} phonon at the Γ point is softened due to electronic screening (a Kohn anomaly). For photoexcited graphite, the screening is weakened because the excited electrons near the E_F cannot follow the fast in-plane nuclear motion. Thus, the non-adiabatic electron–phonon decoupling leads to the stiffening of the E_{2g2} phonon without changing the bond length.

The coherent interlayer shear (E_{2g1}) phonon has a frequency as low as 1.3 THz (42 cm^{-1}), which makes the phonon mode a difficult target for CW Raman scattering but ideal for the time-resolved measurements. The frequency of the E_{2g1} phonon showed a red-shift upon heating to room temperature, but the dephasing rate had no significant temperature dependence [52]. The latter result suggested that the dephasing of this low-frequency phonon is dominated by coupling with electrons and/or defects. Under intense photoexcitation, the E_{2g1} phonon exhibited an amplitude collapse-revival (Fig. 2.14) [53], which looked similar to the collapse-revival in Bi and Sb (Fig. 2.10).

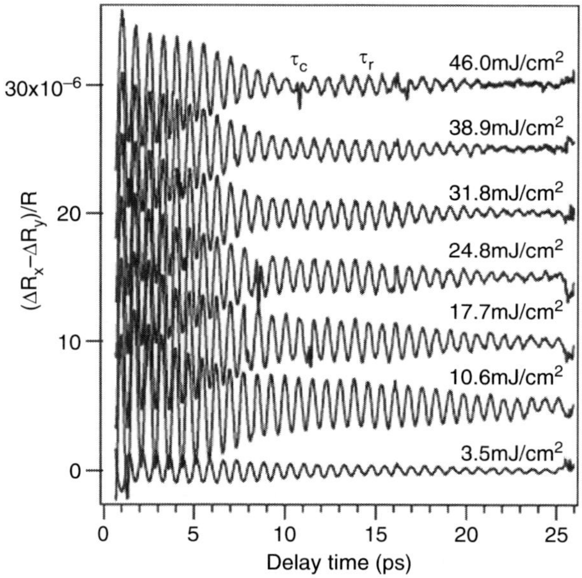

Fig. 2.14. Oscillatory part of the anisotropic reflectivity change of graphite under intense photoexcitation with 140 fs pulses at 1.55 eV at 7 K. From [53]

2.5.3 Exciton–Phonon and Phonon–Phonon Couplings in Carbon Nanotubes

With a structure of rolled up graphene sheets, carbon nanotubes (CNTs) have much in common in their electronic and phonon properties with those of graphite, for example, the linearly dispersing band structure near E_F and the Raman G peak at $\sim$1,600 cm^{-1}. Yet, extensive experimental and theoretical studies have found clear evidence of quantum size effects in CNTs. CNTs sustain the Raman active radial breathing modes (RBMs), whose frequency gives a good measure of their diameter. The tangential modes, which contribute to the Raman G peak, can split into different symmetries depending on the chirality.

Transient transmittance of single-walled carbon nanotubes (SWNTs) in suspension was modulated at two periods of $\sim$140 and 21 fs, corresponding to the RBM and G mode, respectively [54, 55]. The amplitude and the frequency of the coherent RBMs exhibited a clear excitation-wavelength dependence (Fig. 2.15) [54]. The different frequencies were attributed to SWNTs with different diameters coming to the excitonic resonance. The FT spectra of the coherent RBMs in Fig. 2.15 had noticeable differences from the resonant Raman spectra, such as the different intensities and better frequency resolution.

The most striking feature of the coherent G mode was the periodic modulation in its frequency (Fig. 2.16) [55]. As the modulation period coincided

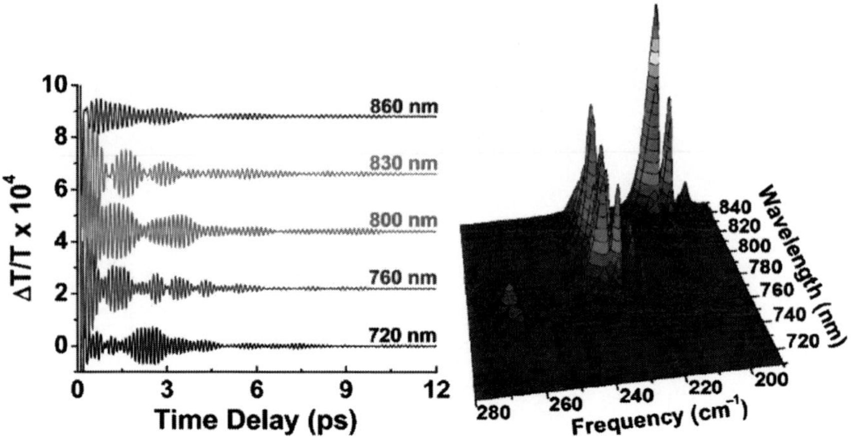

Fig. 2.15. *Left*: oscillatory part of the transient transmittance of SWNTs in a suspension excited and measured with 50 fs pulses at different photon energies. *Right*: a 3D plot of its FT spectrum obtained over a photon energy range of 1.746–1.459 eV (wavelength of 710–850 nm). From [54]

with that of the RBM, a straightforward interpretation was given in terms of the anharmonic coupling between the G mode and RBM on the excited excitonic state. The FT spectrum of the transient transmission signal gave almost equally spaced sidebands around the relatively sharp G peak at 1,588 cm^{-1}, while the Raman spectrum of the same sample consisted of broad G and D bands. Quantum chemical modeling showed that the frequency modulation was due to a corrugation of the SWNT surface on photoexcitation, leading to a coupling between the longitudinal and radial vibrations in the excited state.

2.6 Coherent Optical Phonons in Metals

When metals have Raman active phonons, optical pump-probe techniques can be applied to study their coherent dynamics. Hase and coworkers observed a periodic oscillation in the reflectivity of Zn and Cd due to the coherent E_{2g} phonons (Fig. 2.17) [56]. The amplitude of the coherent phonons of Zn decreased with raising temperature, in accordance with the photoinduced quasi-particle density n_p, which is proportional to the difference in the electronic temperature before and after the photoexcitation (Fig. 2.17). The result indicated the resonant nature of the ISRS generation of coherent phonons. Under intense (mJ/ cm^2) photoexcitation, the coherent E_g phonons of Zn exhibited a transient frequency shift similar to that of Bi (Fig. 2.9), which can be understood as the Fano interference [57]. A transient frequency shift was aslo observed for the coherent transverse optical (TO) phonon in polycrystalline Zr film, in spite of much weaker photoexcitation [58].

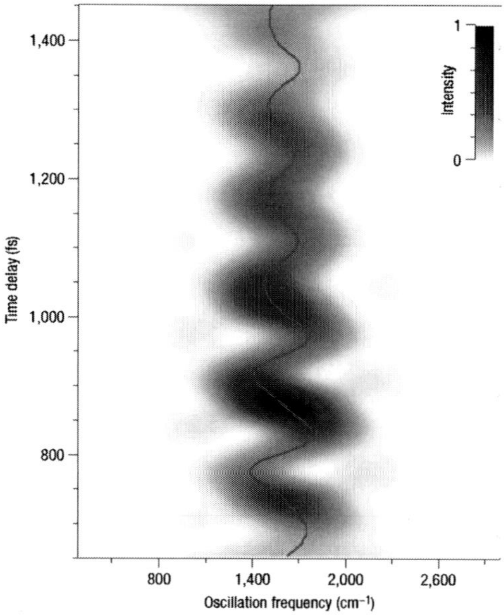

Fig. 2.16. G-mode frequency of SWNTs as a function of pump-probe time delay obtained from transient transmission measurement using a sub-10 fs pulse at 2.1 eV. From [55]

Coherent optical phonons from ferromagnetic systems attract interests in the context of ultrafast magnetic switching. SH signals from Gd(0001) surface in a magnetic field presented a previously unknown phonon–magnon coupled mode at 3 THz under resonant excitation of majority and minority spin subsystems (Fig. 2.18) [32, 59–61]. The frequency of the oscillation suggested the Γ_{3+} interlayer phonon mode at the surface. The vibration modifies the interaction of the spins in the adjacent layers significantly, and thus leads to the observed periodic modulation of the magnetization at the phonon frequency [60]. Above the Curie temperature T_C, the amplitudes of the coherent phonons of Gd(0001) and Tb(0001) did not depend on temperature significantly (Fig. 2.19), and the coherent phonon generation was attributed to DECP as introduced in Sect. 2.2.2. Below T_C, the amplitudes increased remarkably with a dependence that resembled the magnetization itself, indicating the coherent phonon excitation related to the ferromagnetic order [59]. One can consider a vibrational potential depending on the interlayer distance through the exchange interaction (spin) as well as the electron density (charge). If photo-induced spin-flip contributes to an additional shift in the potential minimum, the coherent phonons can be induced displacively by the spins. The damping rate of the coherent phonons for Gd and Tb increased monotonically with temperature, but close to T_C it was strongly reduced [59, 62]. The result indicated that phonon–magnon scattering dominates the

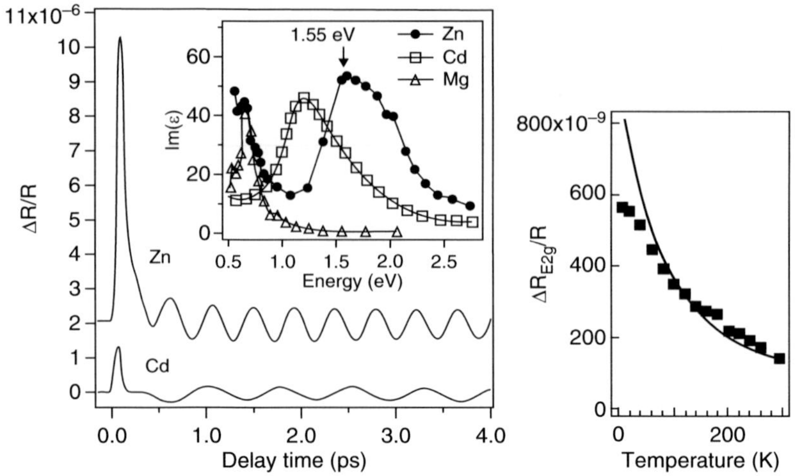

Fig. 2.17. *Left*: transient reflectivity change of Zn and Cd at 7 K. Inset shows the imaginary part of the dielectric function of Zn, Cd, and Mg. *Right*: amplitude of the coherent E_{2g} phonon of Zn as a function of temperature. *Solid curve* in the right panel represents the fit to n_{p}. From [56]

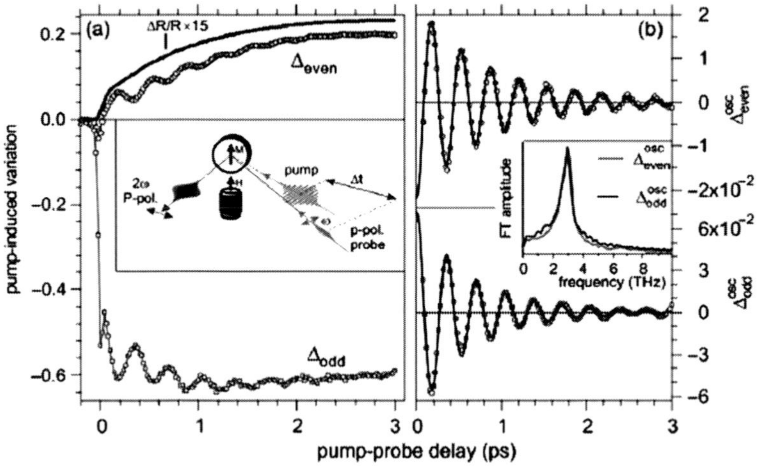

Fig. 2.18. (**a**) The even (*upper panel*) and odd (*lower panel*) SH responses of a 20 nm Gd(0001) film at 90 K using 815 nm/35 fs laser pulses. Transient reflectivity change is also displayed in the upper panel (*solid black curve*). The inset shows the experimental scheme with the magnetization oriented perpendicular to the plane of incidence. (**b**) The oscillatory part of the even and odd SH fields extracted from (**a**). The inset shows the corresponding FT spectra. From [59]

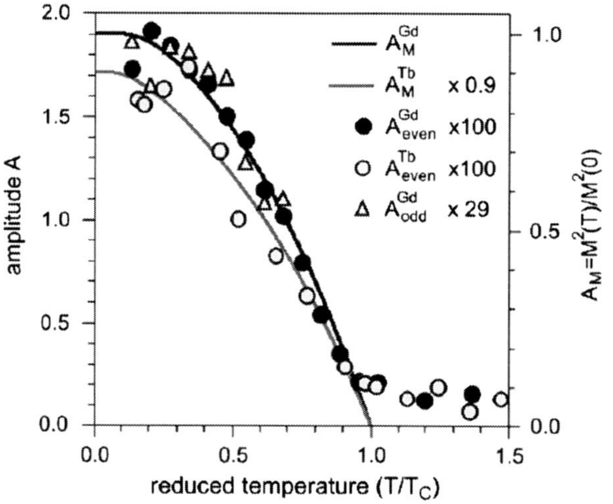

Fig. 2.19. Temperature dependence of the amplitudes of coherent phonons of Gd(0001) and Tb(0001). On the right axis, A_M^{Gd} and A_M^{Tb} show the square of calculated spontaneous magnetization given by the Brillouin function with $J_{\mathrm{Gd}}=7/2$ and $J_{\mathrm{Tb}}=6/2$ representing the magnetic moment of 4f electrons. From [59]

damping of the coherent phonons below T_{C}, together with phonon–phonon scattering and scattering by hot electrons.

2.7 Coherent Phonons in Other Materials

Coherent optical phonons and LO phonon–plasmon coupled (LOPC) modes have been investigated extensively in bulk III-V semiconductors, as well as on their surfaces and in their heterostructures. References [9,10,14] give comprehensive reviews of the experimental achievements, together with theoretical descriptions. Recently, Chang and coworkers confirmed that plasmons couple with both A_1(LO) E_1(LO) phonons to form LOPC modes in wurzite InN, and estimated the electron effective mass from the LOPC frequency [63, 64]. Hase estimated the electron mobility in GaAs by making use of the carrier density-dependence of the LOPC frequency [65]. Coherent LO phonons are investigated also in II–VI semiconductors and their nanostructures [66–71]. Below-bandgap excitation on ZnTe [66] and ZnSe [68] induces coherent phonons via ISRS, while TDFS dominates the coherent phonon generation for above-gap excitation on CdTe [69].

Recent development of ultrashort intense laser pulses has enabled the observation of small-amplitude, high-frequency phonons in wide-gap materials. Typical examples include diamond (Sect. 2.5.1), GaN [72], ZnO [73, 74], and TiO_2 [75,76]. Onishi and coworkers observed the bulk and surface phonon modes of TiO_2 at four different frequencies in their TRSHG measurements

[75, 76]. When the surface was covered with trimethyl acetate (TMA), which has no resonance with the pump and probe wavelengths, no molecular vibration was observed. When adsorbates were replaced by *p*-nitrobenzoate (*p*NB), which has two-photon resonance, the SH intensity showed a modulation at a fifth frequency assigned as a molecular vibration [76]. The study thus provided a first successful application of TRSHG to organic adsorbates.

Systematic TRSHG studies on alkali-atom adsorbed metal surfaces by Matsumoto and coworkers provided a deep insight on how coherent motions are created under very different electronic configurations [15, 77, 78]. The results showed that the coherent phonon generation critically depends on the surface and bulk electronic structure of the substrate.

Coherent optical phonons can couple with localized excitations such as excitons and defect centers. For example, strong exciton–phonon coupling was demonstrated for lead phtalocyanine (PbPc) [79] and CuI [80] as an intense enhancement of the coherent phonon amplitude at the excitonic resonances. In alkali halides [81–83], nuclear wave-packets localized near F centers were observed as periodic modulations of the luminescence spectra.

Coherent phonons and phonon–polaritons were utilized not only to monitor the temperature-driven phase transitions [7,8,84–87] but also to investigate the photo-induced phase transitions (PIPT) in the strongly correlated systems in the real time. Photodoping of holes into VO_2 induced the insulator-to-metal transition, which exhibited a delay of several tens of fs associated with the phonon connecting the two crystallographic phases [88]. Optical measurements on perovskite manganite $Pr_{0.7}Ca_{0.3}MnO_3$ revealed a 31 THz oscillation that was suggestive of coherent orbital wave [89]. For one-dimensional extended Peierls–Hubbard system $[Pd(chxn)_2Br]Br_2$, photo-induced formation of Mott–Hubbard state was observed, followed by a coherent oscillation of the bridging-Br [90]. Organic conductors such as TTF-CA [91–93], K-TCNQ [94,95], (EDO-TTF)$_2$PF$_6$ [96–98], and BEDT-TTF salt [99] are specifically under extensive investigation. Coherent oscillations due to dimerization of the molecules were observed during the photo-induced ionic-to-neutral and neutral-to-ionic transitions in TTF-CA, and was associated to the spin-Peierls instability [91, 92]. (EDO-TTF)$_2$PF$_6$ exhibited a nonequilibrium charge-disproportionate state upon photoexcitation, and the coherent intermolecular bending mode was observed at a frequency depending on the probe wavelength [96–98].

In a heavy fermion compound $Yb_{14}MnSb_{11}$, the dephasing rate of the coherent optical phonons decreased with lowering temperature above Curie temperature T_C, but increased below T_C. The results were attributed to the coupling between an optical phonon mode and the Kondo effect [100].

2.8 Concluding Remarks

In this chapter, we have reviewed the recent experimental researches on the coherent optical phonons using optical detection technique. Optical detection has great advantages in its experimental simplicity, higher time-resolution and higher signal-to-noise ratio. The disadvantage is that the interpretation of the experimentally obtained quantity is not straightforward. Higher signal-to-noise ratio implies that it can probe lattice dynamics under a weak optical perturbation as well as an intense one. The extremely nonequilibrium phenomena under intense perturbation, including PIPT, are attractive in their own right. However, they are challenging targets because they are often associated with irreversible processes – which pump-probe scheme is not designed for. Even in group V semimetals, which are the model materials of coherent phonon studies, the lattice dynamics under intense excitation still lack a comprehensive understanding. We expect recent rapid developments in theoretical researches, in addition to systematic experimental studies, contribute to fully reveal the nontequilibrium phononic and electronic dynamics in solids.

References

1. S. De Silvestri, J.G. Fujimoto, E.P. Ippen, E.B. Gamble, L.R. Williams, K.A. Nelson, Chem. Phys. Lett. **116**(2–3), 146 (1985)
2. T.K. Cheng, S.D. Brorson, A.S. Kazeroonian, J.S. Moodera, G. Dresselhaus, M.S. Dresselhaus, E.P. Ippen, Appl. Phys. Lett. **57**(10), 1004 (1990)
3. G.C. Cho, W. Kütt, H. Kurz, Phys. Rev. Lett. **65**(6), 764 (1990)
4. J.M. Chwalek, C. Uher, J.F. Whitaker, G.A. Mourou, J.A. Agostineli, M. Lelental, Appl. Phys. Lett. **57**(16), 1696 (1990)
5. T. Dekorsy, W. Kutt, T. Pfeifer, H. Kurz, Europhys. Lett. **23**(3), 223 (1993)
6. M. Hase, K. Mizoguchi, H. Harima, S. Nakashima, M. Tani, K. Sakai, M. Hangyo, Appl. Phys. Lett. **69**(17), 2474 (1996)
7. R. Merlin, Solid State Commun. **102**(2–3), 207 (1997)
8. H.J. Bakker, S. Hunsche, H. Kurz, Rev. Modern Phys. **70**(2), 523 (1998)
9. T. Dekorsy, G.C. Cho, H. Kurz, in *Coherent Phonons in Condensed Media* ed. by M.Cardona, G. Güntherodt. Light Scattering in Solids VIII (Springer, Berlin, 2000), p. 169
10. M. Först, T. Dekorsy, *Coherent Phonons in Bulk and Low-Dimensional Semiconductors*, ed. by S. De Silvestri, G. Cerullo, G. Lanzani. Coherent Vibrational Dynamics (CRC, Boca Raton, 2007), p. 129
11. G. Lanzani, in *Coherent Phonon Dynamics in π-Conjugated Chains*, ed. by S. De Silvestri, G. Cerullo, G. Lanzani. Coherent Vibrational Dynamics (CRC, Boca Raton, 2007), p. 93
12. M. Gühr, in *Coherent Dynamics of Halogen Molecules in Rare Gas Solids*, ed. by S. De Silvestri, G. Cerullo, G. Lanzani. Coherent Vibrational Dynamics (CRC, Boca Raton, 2007), p. 173
13. S.L. Dexheimer, in *Coherent Vibrational Dynamics of Exciton Self-Trapping in Quasi-One-Dimensional Systems*, ed. by S. De Silvestri, G. Cerullo, G. Lanzani. Coherent Vibrational Dynamics (CRC, Boca Raton, 2007), p. 223
14. Y.-M. Chang, L. Xu, H.W.K. Tom, Chem. Phys. **251**(1–3), 283 (2000)
15. Y. Matsumoto, K. Watanabe, Chem. Rev. **106**(10), 4234 (2006)
16. A.V. Kuznetsov, C.J. Stanton, Phys. Rev. Lett. **73**(24), 3243 (1994)

17. T. Dekorsy, H. Auer, C. Waschke, H.J. Bakker, H.G. Roskos, H. Kurz, Phys. Rev. Lett. **74**(5), 738 (1995)
18. M. Tani, R. Fukasawa, H. Abe, S. Matsuura, K. Sakai, S. Nakashima, J. Appl. Phys. **83**(5), 2473 (1998)
19. L. Dhar, J. Rogers, K.A. Nelson, Chem. Rev. **94**, 157 (1994)
20. K.J. Yee, Y.S. Lim, T. Dekorsy, D.S. Kim, Phys. Rev. Lett. **86**(8), 1630 (2001)
21. K. Ishioka, M. Kitajima, O.V. Misochko, J. Appl. Phys. **103**, 123505 (2008)
22. Y.-M. Chang, Appl. Phys. Lett. **82**(11), 1781 (2003)
23. E.S. Zijlstra, L.L. Tatarinova, M.E. Garcia, Phys. Rev. B **74**(22), 220301 (2006)
24. T.K. Cheng, J. Vidal, H.J. Zeiger, G. Dresselhaus, M.S. Dresselhaus, E.P. Ippen, Appl. Phys. Lett. **59**(16), 1923 (1991)
25. K. Ishioka, M. Kitajima, O.V. Misochko, J. Appl. Phys. **100**, 093501 (2006)
26. G.A. Garrett, T.F. Albrecht, J.F. Whitaker, R. Merlin, Phys. Rev. Lett. **77**(17), 3661 (1996)
27. H.J. Zeiger, J. Vidal, T.K. Cheng, E.P. Ippen, G. Dresselhaus, M.S. Dresselhaus, Phys. Rev. B **45**(2), 768 (1992)
28. T.E. Stevens, J. Kuhl, R. Merlin, Phys. Rev. B **65**, 144304 (2002)
29. D.M. Riffe, A.J. Sabbah, Phys. Rev. B **76**(8), 085207 (2007)
30. T. Dekorsy, T. Pfeifer, W. Kütt, H. Kurz, Phys. Rev. B **47**(7), 3842 (1993)
31. G.C. Cho, H.J. Bakker, T. Dekorsy, H. Kurz, Phys. Rev. B **53**(11), 6904 (1996)
32. U. Bovensiepen, J. Phys. Condens. Matter **19**, 235417 (2007)
33. L. Min, R.J.D. Miller, Appl. Phys. Lett. **56**(6), 524 (1990)
34. Y.-M. Chang, L. Xu, H.W.K. Tom, Phys. Rev. Lett. **78**(24), 4649 (1997)
35. K. Sokolowski-Tinten, C. Blome, J. Blums, A. Cavalleri, C. Dietrich, A. Tarasevitch, I. Uschmann, E. Förster, M. Kammler, M. Horn von Hoegen, D. von der Linde, Nature **422**(6929), 287 (2003)
36. M. Hase, M. Kitajima, S. Nakashima, K. Mizoguchi Phys. Rev. Lett. **88**(6), 067401 (2002)
37. M.F. DeCamp, D.A. Reis, P.H. Bucksbaum, R. Merlin, Phys. Rev. B **64**, 092301 (2001)
38. O.V. Misochko, K. Ishioka, M. Hase, M. Kitajima, J. Phys. Condens. Matter **19**(15), 156227 (2007)
39. E.D. Murray, D.M. Fritz, J.K. Wahlstrand, S. Fahy, D.A. Reis, Phys. Rev. B **72**, 060301(R) (2005)
40. E.D. Murray, S. Fahy, D. Prendergast, T. Ogitsu, D.M. Fritz, D.A. Reis, Phys. Rev. B **75**(18), 184301 (2007)
41. T. Dumitrica, J. Phys. Condens. Matter **19**, 181003 (2007)
42. O.V. Misochko, K. Ishioka, M. Hase, M. Kitajima, J. Phys. Condens. Matter **18**, 10571 (2006)
43. O.V. Misochko, M. Hase, K. Ishioka, M. Kitajima, Phys. Rev. Lett. **92**(19), 197401 (2004)
44. T. Baumert, M. Grosser, R. Thalweiser, G. Gerber, Phys. Rev. Lett. **67**(27), 3753 (1991)
45. M.J.J. Vrakking, D.M. Villeneuve, A. Stolow, Phys. Rev. A **54**(1), R37 (1996)
46. M.S. Diakhate, E.S. Zijlstra, M.E. Garcia, Appl. Phys. A **96**, 5177 (2009)
47. M. Hase, M. Kitajima, A.M. Constantinescu, H. Petek, Nature **426**, 51 (2003)
48. T. Pfeifer, W. Kütt, H. Kurz, R. Scholz, Phys. Rev. Lett. **69**(22), 3248 (1992)
49. M. Nakajima, H. Harima, K. Morita, K.M. Itoh, K. Mizoguchi, Phys. Rev. B **63**, 161304(R) (2001)

50. K. Ishioka, M. Hase, M. Kitajima, H. Petek, Appl. Phys. Lett. **89**, 231916 (2006)
51. K. Ishioka, M. Hase, M. Kitajima, L. Wirtz, A. Rubio, H. Petek, Phys. Rev. B **77**(12), 121402 (2008)
52. T. Mishina, K. Nitta, Y. Masumoto, Phys. Rev. B **62**(41), 2908 (2000)
53. R. Lu, K. Ishioka, O.V. Misochko, M. Hase, M. Kitajima, Surf. Sci. **593**(1–3), 116 (2005)
54. Y.-S. Lim, K.-J. Yee, J.-H. Kim, E.H. Haroz, J. Shaver, J. Kono, S.K. Doorn, R.H. Hauge, R.E. Smalley, Nano Lett. **6**(12), 2696 (2006)
55. A. Gambetta, C. Manzoni, E. Menna, M. Meneghetti, G. Cerullo, G. Lanzani, S. Tretiak, A. Piryatinski, A. Saxena, R.L. Martin, A.R. Bishop, Nat. Phys. **2**, 515 (2006)
56. M. Hase, K. Ishioka, J. Demsar, K. Ushida, M. Kitajima, Phys. Rev. B **71**(18), 184301 (2005)
57. M. Hase, J. Demsar, M. Kitajima, Phys. Rev. B **74**(21), 212301 (2006)
58. V.V. Kruglyak, R.J. Hicken, G.P. Srivastava, M. Ali, B.J. Hickey, A.T.G. Pym, B.K. Tanner, Phys. Rev. B **76**(1), 012301 (2007)
59. A. Melnikov, I. Radu, A. Povolotskiy, T. Wehling, A. Lichtenstein, U. Bovensiepen, J. Phys. D Appl. Phys. **41**(16), 164004 (2008)
60. A. Melnikov, I. Radu, U. Bovensiepen, O. Krupin, K. Starke, E. Matthias, M. Wolf, Phys. Rev. Lett. **91**, 227403 (2003)
61. U. Bovensiepen, A. Melnikov, I. Radu, O. Krupin, K. Starke, M. Wolf, E. Matthias, Phys. Rev. B **69**(23), 235417 (2004)
62. A. Melnikov, A. Povolotskiy, U. Bovensiepen, Phys. Revi. Lett. **100**(24), 247401 (2008)
63. Y.M. Chang, H.W. Chu, C.H. Shen, H.Y. Chen, S. Gwo, Appl. Phys. Lett. **90**(7), 72111 (2007)
64. Y.M. Chang, H.W. Lin, Y.L. Hong, S. Gwo, J. Appl. Phys. **104**(3), 036105 (2008)
65. M. Hase, Appl. Phys. Lett. **94**(11), 112111 (2009)
66. Y.-S. Lim, S.-C. Yoon, K.-J. Yee, J.-H. Lee, D.S. Kim, D. Lee, Phys. Rev. B **68**(15), 153308 (2003)
67. J.K. Wahlstrand, T.E. Stevens, J. Kuhl, R. Merlin, Physica B **(316–317)**, 55 (2002)
68. Y.-S. Lim, S.-C. Yoon, K.-J. Yee, Y.-H. Ahn, E. Oh, J.-H. Lee, Appl. Phys. Lett. **82**(15), 2446 (2003)
69. K. Ishioka, M. Kitajima, J. Irisawa, Y. Hironaka, K. Ushida, K.G. Nakamura, Jpn. J. Appl. Phys. **45**(12), 9111 (2006)
70. A.V. Bragas, C. Aku-Leh, S. Costantino, A. Ingale, J. Zhao, R. Merlin, Phys. Rev. B **69**(20), 205306 (2004)
71. S. Tripathy, H.P. Wagner, A. Ueta, D. Hommel, Phys. Rev. B **75**(24), 245316 (2007)
72. K.J. Yee, K.G. Lee, E. Oh, D.S. Kim, Phys. Rev. Lett. **88**(10), 105501 (2002)
73. I.H. Lee, K.J. Yee, K.G. Lee, E. Oh, D.S. Kim, Y.S. Lim, J. Appl. Phys. **93**(8), 4939 (2003)
74. C. Aku-Leh, J. Zhao, R. Merlin, J. Menendez, M. Cardona, Phys. Rev. B **71**(20), 205211 (2005)
75. S. Fujiyoshi, T. Ishibashi, H. Onishi, J. Phys. Chem. B **109**(18), 8557 (2005)
76. T. Nomoto, H. Onishi, Chem. Phys. Lett. **455**(4–6), 343 (2008)

77. K. Watanabe, N. Takagi, Y. Matsumoto, Phys. Rev. B **71**, 085414 (2005)
78. M. Fuyuki, K. Watanabe, D. Ino, H. Petek, Y. Matsumoto, Phys. Rev. B **76**(11), 115427 (2007)
79. K. Mizoguchi, S. Fujita, M. Nakayama, Appl. Phys. A **78**(4), 461 (2004)
80. O. Kojima, K. Mizoguchi, M. Nakayama, J. Lumin. **112**(1–4), 80 (2005)
81. T. Koyama, Y. Takahashi, M. Nakajima, T. Suemoto, Phys. Rev. B **73**(16), 161102 (2006)
82. T. Koyama, Y. Takahashi, M. Nakajima, T. Suemoto, J. Chem. Phys. **124**(22), 221104 (2006)
83. T. Koyama, Y. Takahashi, M. Nakajima, T. Suemoto, Phys. Rev. B **76**, 115122 (2007)
84. T. Kohmoto, K. Tada, T. Moriyasu, Y. Fukuda, Phys. Rev. B **74**(6), 64303 (2006)
85. O.V. Misochko, M. Tani, K. Sakai, K. Kisoda, S. Nakashima, V.N. Andreev, F.A. Chudnovsky, Phys. Rev. B **58**(19), 12789 (1998)
86. H.T. Kim, Y.W. Lee, B.J. Kim, B.G. Chae, S.J. Yun, K.Y. Kang, K.J. Han, K.J. Yee, Y.S. Lim, Phys. Rev. Lett. **97**(26), 266401 (2006)
87. H. Kamioka, S. Saito, M. Isobe, Y. Ueda, T. Suemoto, Phys. Rev. Lett. **88**(12), 127201/1 (2002)
88. A. Cavalleri, T. Dekorsy, H.H.W. Chong, J.C. Kieffer, R.W. Schoenlein. Phys. Rev. B **70**(16), 161102 (2004)
89. D. Polli, M. Rini, S. Wall, R.W. Schoenlein, Y. Tomioka, Y. Tokura, G. Cerullo, A. Cavalleri, Nat. Mater. **6**, 643 (2007)
90. H. Matsuzaki, M. Yamashita, H. Okamoto, J. Phys. Soc. Jpn. **75**(12), 123701/1 (2006)
91. S. Iwai, S. Tanaka, K. Fujinuma, H. Kishida, H. Okamoto, Y. Tokura, Phys. Rev. Lett. **88**(5), 057402 (2002)
92. H. Okamoto, Y. Ishige, S. Tanaka, H. Kishida, S. Iwai, Y. Tokura, Phys. Rev. B **70**(16), 165202 (2004)
93. K. Tanimura, Phys. Rev. B **70**(14), 144112 (2004)
94. H. Okamoto, K. Ikegami, T. Wakabayashi, Y. Ishige, J. Togo, H. Kishida, H. Matsuzaki, Phys. Rev. Lett. **96**(3), 037405 (2006)
95. K. Ikegami, K. Ono, J. Togo, T. Wakabayashi, Y. Ishige, H. Matsuzaki, H. Kishida, H. Okamoto, Phys. Rev. B **76**(8), 085106 (2007)
96. M. Chollet, L. Guerin, N. Uchida, S. Fukaya, H. Shimoda, T. Ishikawa, K. Matsuda, T. Hasegawa, A. Ota, H. Yamochi, G. Saito, R. Tazaki, S. Adachi, S. Koshihara, Science **307**(5706), 86 (2005)
97. K. Onda, S. Ogihara, T. Ishikawa, Y. Okimoto, X. Shao, H. Yamochi, G. Saito, S. Koshihara, J. Phys. Condens. Matter **20**(22), 224018 (2008)
98. K. Onda, S. Ogihara, K. Yonemitsu, N. Maeshima, T. Ishikawa, Y. Okimoto, X. Shao, Y. Nakano, H. Yamochi, G. Saito, S. Koshihara, Phys. Rev. Lett. **101**(6), 067403 (2008)
99. S. Iwai, K. Yamamoto, A. Kashiwazaki, F. Hiramatsu, H. Nakaya, Y. Kawakami, K. Yakushi, H. Okamoto, H. Mori, Y. Nishio, Phys. Rev. Lett. **98**(9), 097402 (2007)
100. K.S. Burch, E.M.C. Elbert, D. Talbayev, B.C. Sales, D. Mandrus, A.J. Taylor, R.D. Averitt, Phys. Rev. Lett. **100**(2), 026409 (2008)

3

Coherent Lattice Oscillations in Solids and Their Optical Control Part II. New Detection Techniques and Optical Control

Kunie Ishioka and Oleg V. Misochko

Summary. Although optical detection is still most widely used in the experimental studies of coherent optical phonons, new detection techniques such as X-ray diffraction, THz, and photoemission spectroscopies are providing more and more important insight, especially in crystals with strong electron–phonon interaction. Making use of the classical interference, shaped optical pulses have been employed to control the amplitude of coherent phonons and to selectively excite a single phonon mode out of many. Experiments with multiple intense pulses have indicated a possibility of driving the crystal into an extremely nonequilibrium vibrational state that a single pulse cannot achieve.

3.1 Coherent Phonons Detected by Novel Techniques

3.1.1 Time Resolved X-Ray Diffraction

Time-resolved X-ray diffraction (TRXRD), illustrated in Fig. 3.1, provides a powerful technique to probe directly the structural dynamics of crystals far from the equilibrium. It employs visible pump pulses from a laser, and laser- or accelerator-based X-ray probe pulses [1, 3]. As X-ray diffraction can in principle probe $k \neq 0$ phonons, TRXRD has the potential to reveal the energy transfer dynamics, for example, from the zone-center to the zone-boundary phonons.

TRXRD detects the propagation of coherent *acoustic* phonons as a transient change in the diffraction angles. In contrast, the atomic motions associated with coherent *optical* phonons modify only the Bragg peak intensity, because they do not change the barycentric positions of the crystal lattice. The Bragg peak intensity is proportional to the squared modulus of the structure factor [1, 3, 4]:

$$S(h,k,l,t) = S_0 - i\Big[\sum_{j=1}^{N} f_j \boldsymbol{G}_j \boldsymbol{u}_j \exp(-\mathrm{i}\boldsymbol{G}_j \boldsymbol{r}_j)\Big] \sin(\omega t). \tag{3.1}$$

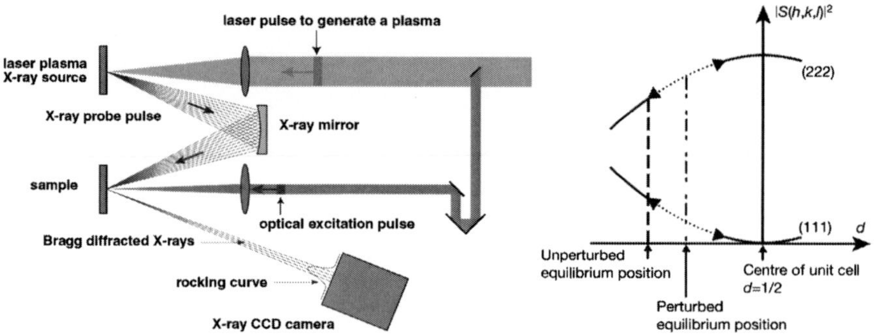

Fig. 3.1. *Left*: visible pump/X-ray probe scheme for femtosecond TRXRD experiments. Hard X-ray pulses are generated by shining intense femtosecond laser pulses on a metal target (laser plasma X-ray source). *Right*: geometrical structure factor of bismuth as a function of inter-atomic distance for diffraction from (111) and (222) lattice planes. From [1] and [2]

If $|S(h,k,l)|^2$ can be approximated by a linear function of displacement ($\Delta \boldsymbol{r}_j$) in the vicinity of the equilibrium position (Fig. 3.1), the modulation in the Bragg peak intensity is a direct measure for nuclear displacement. The quantitative determination of the Bragg intensity requires a high signal-to-noise-ratio detection with high-brightness X-ray pulses. Availability of ultrashort X-ray pulses, now 100 fs or longer, poses another challenge to the detection of coherent optical phonons. In the experiments in which the pump and probe are derived from separate sources (e.g., pump from a laser and probe from an accelerator), synchronization is essential to achieve the high temporal resolution. The difficulty of synchronizing the two sources within a fraction of the probe pulse duration can be circumvented by measuring the relative arrival time on a shot-by-shot basis [5].

The first TRXRD detection of coherent optical phonons was performed by Sokolowski-Tinten and coworkers on a Bi thin film using X-ray pulses generated from laser plasma [2]. As described in Sect. 2.4 the previous chapter, Bi had been studied extensively with optical techniques, yet some controversies remained, for example, on the generation mechanism. Relatively low frequencies of the optical phonons made Bi feasible as a target of the TRXRD experiments. The results, shown in Fig. 3.2, demonstrated that the X-ray intensity from the (111) planes dropped suddenly at $t = 0$ and oscillated around the new equilibrium, confirming the DECP generation of the coherent A_{1g} phonon. The phonon frequency was strongly redshifted to a value (2.12 THz) much lower than the optically observed frequency under a similar excitation density [6]. A following report using an accelerator as a high-brightness X-ray source estimated the shift in the equilibrium position to be 2% of the inter-atomic separation when 1.8% of valence electrons were photoexcited [7]. TRXRD experiments with grazing incidence by Johnson and coworkers revealed the depth dependence of the X-ray signal from Bi

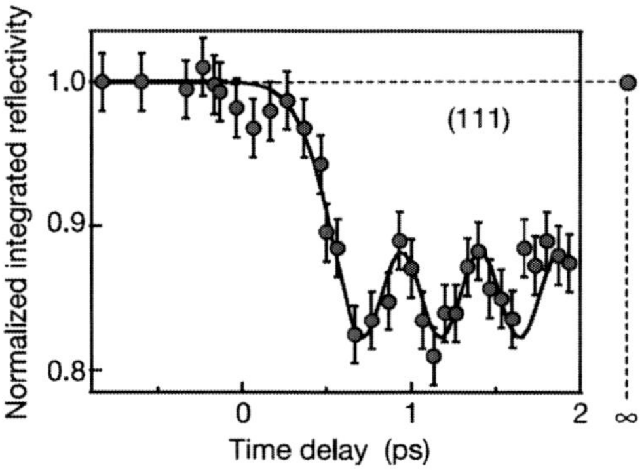

Fig. 3.2. X-ray diffraction efficiency of the (111) reflection of Bi after excitation with an optical pulse at $6\,\mathrm{mJ/cm^2}$. *Circles* and *solid curves* are the experimental data and the fit, respectively. From [2]

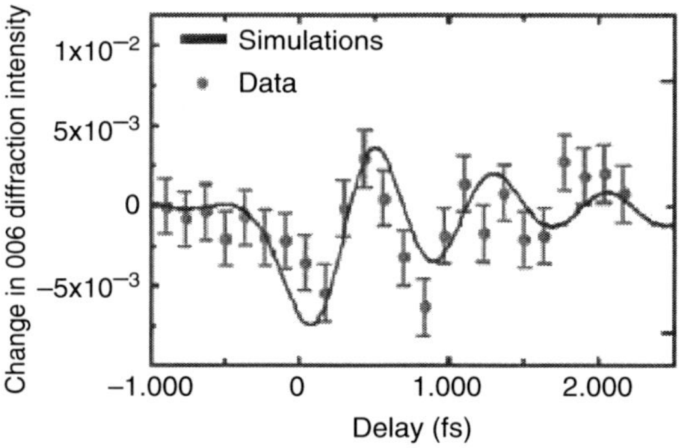

Fig. 3.3. Time-dependent change in the 006 X-ray diffraction intensity of $LiTaO_3$. From [10]

[8]. In combination with DFT calculations, the time- and depth-dependent phonon frequency allows to estimate the effective diffusion rate of $2.3\,\mathrm{cm^2\,s^{-1}}$ and the electron–hole thermalization time of 260 fs for highly excited carriers. A recent experiment by the same group looked at the $(10\bar{1})$ and $(11\bar{2})$ diffractions in search of the coherent E_g phonons. They observed a periodic modulation at 1.3 THz, which was much slower than that expected for the E_g mode, and attributed the oscillation to the squeezed phonon states [9].

One of the applications of TRXRD is to study complex systems where electric fields couple to multiple degrees of freedom. Though femtosecond laser pulses can generate THz radiation from ferroelectric $LiTaO_3$, the corresponding lattice motion remained undetected by optical measurements. Cavalleri and coworkers demonstrated the coherent modulation of the X-ray intensity at 1.5 THz [10], and assigned it as phonon–polariton mode of A_1 symmetry (Fig. 3.3). Nakamura and coworkers detected the coherent LO phonon of CdTe

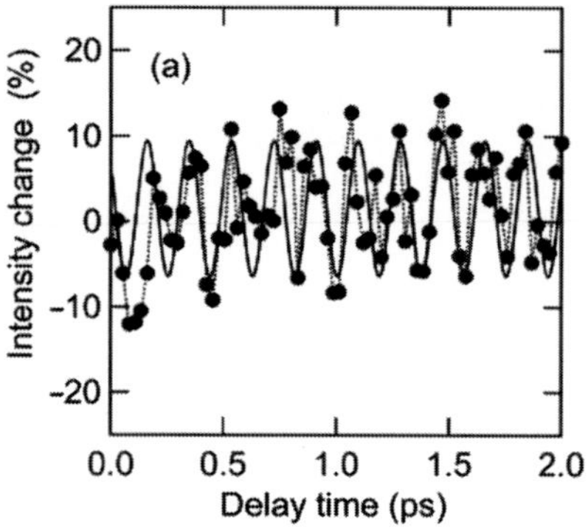

Fig. 3.4. X-ray diffraction intensity of the (111) reflection of CdTe at pump pulse density 0.6 mJ/ cm^2. *Filled circles* and *solid curve* indicate the experimental data and the fit with an oscillation at 5.3 THz, respectively. From [4]

at 5 THz [4] (Fig. 3.4). Though this system had been previously investigated optically, the formation of the LOPC mode left the assignment of the oscillations ambiguous [11]. The real-space displacement due to the coherent LO phonons was estimated to be 0.8% of the distance of lattice planes in the [111] direction.

3.1.2 X-Ray Abrotption Spectroscopy

X-ray absorption spectroscopy (XAS) measures the absorption coefficient as a function of incident X-ray photon energy. For molecules and solids, the absorption spectrum is modulated at energies far above the absorption edge due to the scattering of photoelectrons by neighboring atoms. The interference pattern is Fourier-transformed to yield structural information such as bond length and angles in the framework of extended X-ray absorption fine structure (EXAFS). Time-resolved X-ray absorption spectroscopy (TRXAS) in optical-pump and X-ray probe scheme achieved 20 fs time resolution by using soft X-ray pulses obtained from high harmonic generation. The technique was applied to Si thin film and demonstrated the coherent optical phonons at 17 THz [12].

3.1.3 Time-Resolved THz Spectroscopy

Material response in THz frequency region, which corresponds to far- and mid-infrared electromagnetic spectrum, carries important information for the understanding of both electronic and phononic properties of condensed matter. Time-resolved THz spectroscopy has been applied extensively to investigate the sub-picosecond electron–hole dynamics and the coherent lattice dynamics simultaneously. In a typical experimental setup shown in Fig. 3.5, an

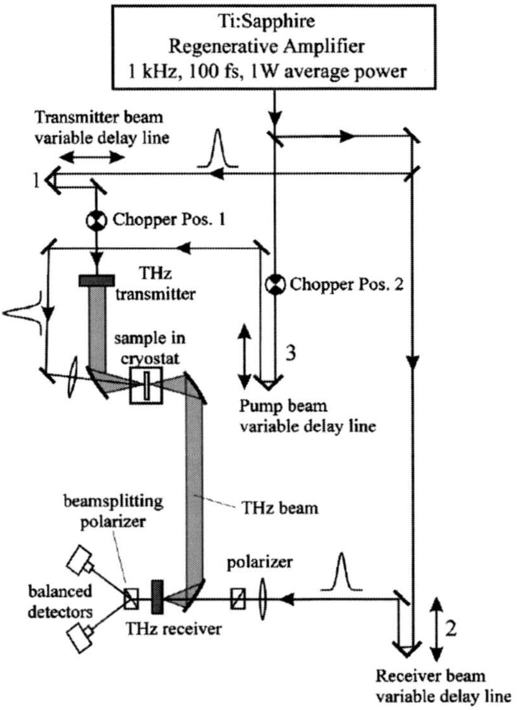

Fig. 3.5. Experimental apparatus for time-resolved THz transmission spectroscopy. The sample is excited with a visible laser pulse delivered by delay line 3. A single-cycle THz electric-field transient probes the polarization response of the sample after time delay t_D scanned by delay line 1. The transmitted THz amplitude is monitored via ultrabroadband electro-optic sampling in a THz receiver as a function of time T scanned by delay line 2. From [13]

intense femtosecond laser pulse is split into three, each of which is responsible for either the THz transmitter, THz receiver, or photoexcitation. The transmitter beam generates an ultrabroadband THz pulse in a nonlinear optical crystal via phase-matched optical rectification [14]. The THz pulse is then transmitted through the sample at time delay t_D with respect to the photoexcitation pulse, and collected onto another nonlinear optical crystal. Phase-matched electro-optic sampling [15] yields both amplitude and phase of the THz probe field by scanning delay T of the receiver beam. The two-dimensional multi-THz scheme provides direct access with femtosecond temporal resolution to the complex-valued ultrabroadband conductivity $\sigma(\omega)$ and, simultaneously, to its pump-induced changes $\Delta\sigma(\omega, t_D)$.

In photoexcited polar semiconductors, the coherent LO phonons couple with photocarriers to form coherent LO phonon–plasmon coupled (LOPC) modes, which exhibit fundamentally different properties from those of bare phonons. Huber and coworkers revealed the ultrafast transition of an optical

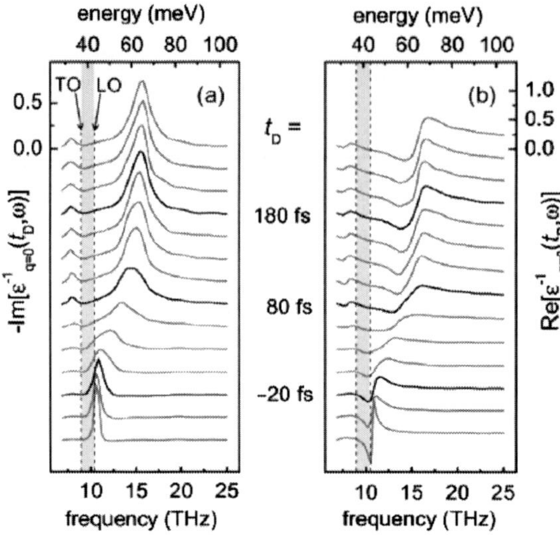

Fig. 3.6. (**a**) Imaginary and (**b**) real parts of the long-wavelength limit of the inverse dielectric function of *i*-InP vs. frequency, showing the ultrafast formation of the LOPC mode after photoexcitation. From [16]

phonon resonance to a coupled phonon–plasmon system in InP [16]. Mutual repulsion and redistribution of the oscillator strength of the interacting phonons and plasmons was seen to emerge within one oscillation cycle of the upper branch of the LOPC modes (Fig. 3.6).

Metal–insulator transitions in strongly correlated systems offer another example on which time-resolved THz transmission spectroscopy can give a fundamental insight. TRXRD [17] and reflectivity [18] measurements on VO_2 had demonstrated the photo-induced phase transition from rutile to monoclinic to occur in sub-picosecond timescale, yet the microscopic mechanism remained unexplored. Time-resolved THz spectroscopy [19] revealed that the lattice polarizability was strongly influenced by the coherent modulation at 6 THz, which is characteristic of impulsively excited wave packet motion of V-V dimers (Fig. 3.7). Depending on the excitation density, the electronic conductivity showed signatures either of excitonic self-trapping or an ultrafast crossover into a metallic phase. The results indicated that the high electronic conductivity is already established after one oscillation cycle of the V-V intradimer mode or in 130 fs, and does not follow the ongoing coherent lattice vibrations.

3.1.4 Time-Resolved Photo-Emission Spectroscopy

Coherent lattice motions can create periodic modulation of the electronic band structure. Time-resolved photo-emission (TRPE) studies [20–22] demonstrated the capability to detect coherent phonons as an oscillatory shift of

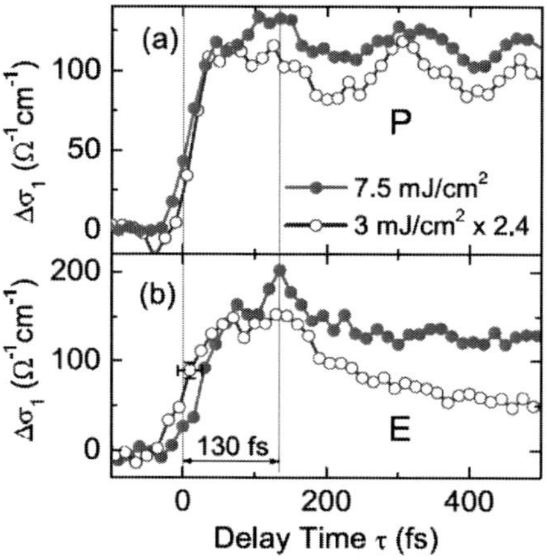

Fig. 3.7. Transient conductivity change of VO_2 for a photon energy of (**a**) 60 meV and (**b**) 100 meV obtained from multi-THz measurements. The traces in (**a**) feature a modulation at 6 THz due to the coherent V-V oscillation. From [19]

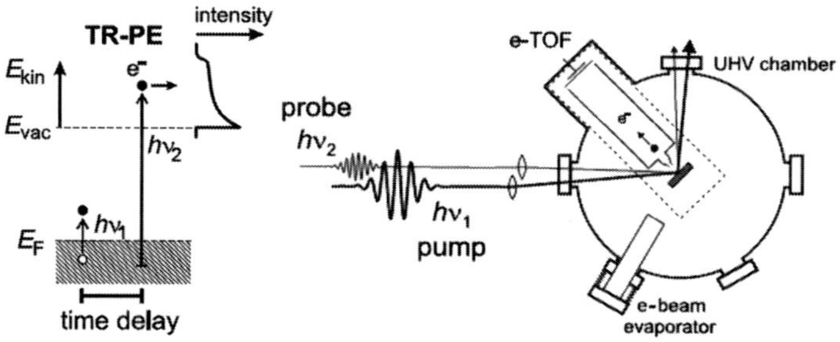

Fig. 3.8. *Left*: schematic illustration of TRPE. The IR pump pulse ($h\nu_1$) perturbs the electronic states of the sample. The photon energy of the UV probe pulse ($h\nu_2$) exceeds the work function and monitors changes in occupied and unoccupied states simultaneously. *Right*: experimental setup for TRPE. Pairs of IR and UV pulses are time delayed with respect to each other and are focused onto the sample surface in the UHV chamber. The kinetic energy of photoelectrons is analyzed by an electron time-of-flight spectrometer (e-TOF). From [23]

the binding energy of an occupied state. In TRPE experiments, the valence band photoemission spectra are obtained using a femtosecond UV probe pulse, and are recorded as a function of delay after an excitation by an IR pump pulse (Fig. 3.8) [23]. When combined with the angle-resolved technique [22], TRPE has a great advantage of capturing the comprehensive picture of the collective excitation through observation of time-, frequency-, and momentum-dependent electronic structure.

Gd(0001) surface, which serves as a model system for a ferromagnetic metal, presented a coherent coupled phonon–magnon mode at 3 THz in the TRSHG measurements, as presented in Sect. 2.6 in the previous chapter.

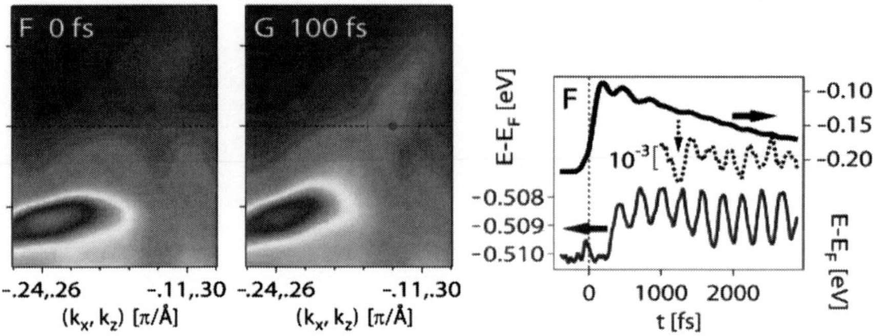

Fig. 3.9. *Left*: photoelectron intensity from $TbTe_3$ surface as a function of energy and momentum for different time delays, showing the ultrafast closing of the CFW gap marked with a dot. *Right*: Time-dependent binding energy of the Te band (*lower trace*) and the CB (*upper trace*), exhibiting a periodic modulation at 2.3 and 3.6 THz, respectively, under strong excitation fluence ($2\,mJ/\,cm^2$). From [22]

The same vibration leads to a periodical shift of the surface-state energy levels via the exchange coupling of the surface spin to the bulk magnetization. Bovensiepen and coworkers observed this oscillation of the binding energy, which with the aid of DFT calculations they translated to the interlayer spacing in picometer scale [20, 23].

Charge density wave (CDW) materials are among the well-established examples of quantum many body problems, and $TbTe_3$ is a model system to study Fermi surface nesting-driven CDW formation. A time- and angle-resolved photoemission spectroscopy (trARPES) study [22] revealed a transient melting of the CDW state (i.e., closing of the CDW gap) to be delayed by 100 fs with respect to the photo-doping (Fig. 3.9). Simultaneously, periodic modulation of the binding energy was observed at different frequencies for different electronic states (2.3 and 3.6 THz for the Te conduction band (CB) and Te band, respectively). Temperature-, pump fluence-, and delay time-dependence of the 2.3 THz oscillation was characteristic of the CDW state, leading to the assignment as the CDW amplitude mode coupled to long-wavelength optical phonons. The amplitude mode oscillation drives the CDW gap modulation. A similar periodic modulation of the binding energy due to the CDW amplitude mode was also reported for TaS_2 [21]. In contrast, the 3.6 THz oscillation of $TbTe_3$ persisted after the melting of CDW, and was attributed to the Te phonon. The results demonstrated that a large enough coherent phonon excitation can drive a transition from an insulating to a metallic state.

3.2 Optical Control of Coherent Phonons

By making use of classical or quantum-mechanical interferences, one can use light to control the temporal evolution of nuclear wavepackets in crystals. An appropriately timed sequence of femtosecond light pulses can selectively excite a vibrational mode. The ultimate goal of such optical control is to prepare an extremely nonequilibrium vibrational state in crystals and to drive it into a novel structural and electromagnetic phase.

3.2.1 Optical Control at Low Excitation Density

The earliest control experiments were performed in double- (or multiple-) pump and probe scheme on optical phonons generated via ISRS in transparent materials by Nelson and coworkers [24, 25]. Shortly later, similar experiments were carried out on coherent phonons generated in semiconductors via TDFS by Dekorsy and coworkers [26], and on those generated in semimetals via DECP by Hase and coworkers [27] (Fig. 2.1 in the previous chapter). These experiments demonstrated that the amplitude of the coherent oscillation can be controlled by varying the temporal separation $\Delta\tau$ between the two pump pulses. $\Delta\tau = nT$ leads to the maximum enhancement of the amplitude with an integer n and the phonon period T, while $\Delta\tau = (n + 1/2)T$ results in complete cancelation.

Within the ISRS framework, the first impulse force kick-starts the coherent nuclear oscillation. At $t = nT$ or $(n+1/2)T$, the nuclei pass their equilibrium position, moving in the direction of the first force or in the opposite, respectively. The second ISRS force then pushes the nuclei to accelerate (Fig. 3.10a) or pulls them to a halt. Such "coherent" control via ISRS was demonstrated

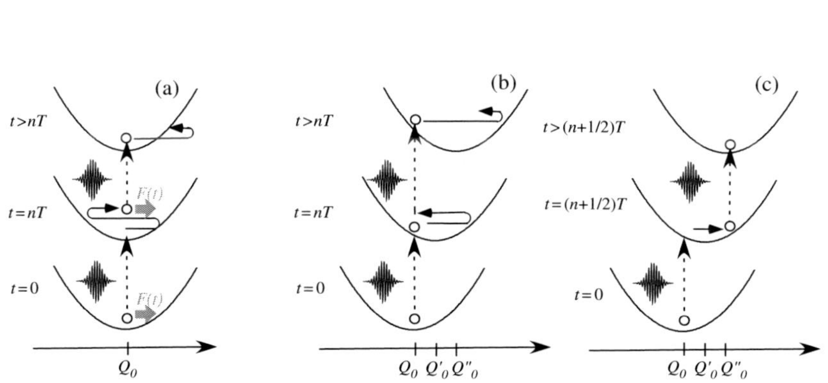

Fig. 3.10. Schematics of interferences of coherent phonons in double-pump experiments. (**a**) constructive interference in ISRS mechanism. *Bold grey arrows* indicate the first and the second ISRS driving forces. (**b**) constructive interferences in DECP mechanism. (**c**) destructive interferences in DECP mechanism

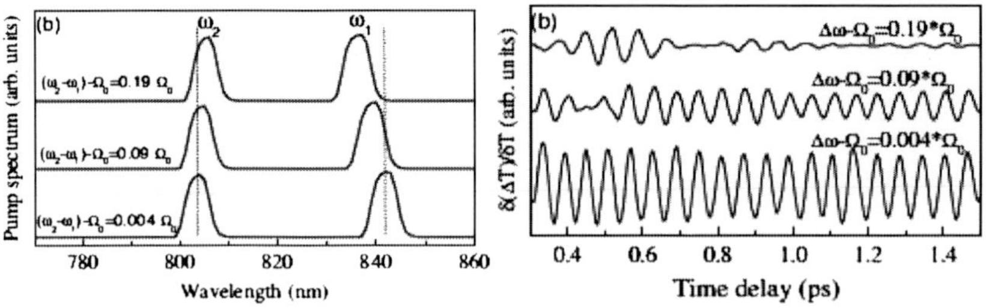

Fig. 3.11. *Left*: spectrum of the tailored pump beam for different spectral separations $\Delta\omega = \omega_2 - \omega_1$ between two spectral packets (ω_1 and ω_2). The bottom case is resonant to the E_2 phonon frequency Ω_0. *Right*: transient transmittance of GaN excited with the tailored pump pulses, showing the enhancement of coherent oscillation of the E_2 phonon for $\omega_1 - \omega_2 \simeq \Omega_0$. From [28]

also for GaN [28] and for zone-folded acoustic phonons in semiconductor MQWs [29–31]. Lee and coworkers demonstrated an enhancement of the E_2 phonon amplitude of GaN by tailoring the spectral separation $(\omega_2 - \omega_1)$ of the shaped pulse to match the phonon frequency Ω_0 (Fig. 3.11) [28]. It should be noted, however, that this result alone does not necessarily prove the ISRS generation, as the spectrum is equivalent to a pulse train with the repetition rate of the difference frequency $\omega_2 - \omega_1 = \Omega_0 = T^{-1}$, which satisfies the integer criterion.

In the framework of DECP, the first pump pulse establishes a new potential surface, on which the nuclei start to move toward the new equilibrium. The nuclei gain momentum and reach the classical turning points of their motion at $t = nT$ and $t = (n + 1/2)T$. The second pump pulse then shifts the equilibrium position, either away from (Fig. 3.10b) or to the current position of the nuclei (Fig. 3.10c). The latter leads to a halt of the nuclear motion. Because photo-excitation of additional electrons can only shift the equilibrium position further in the same direction, the vibrations can only be stopped at their maximum displacement [32].

When the system sustains more than one Raman-active phonon mode, pulse trains can selectively enhance one of the modes and weaken the others by appropriately choosing the repetition rate $\Delta\tau^{-1}$. This idea was effectively demonstrated for different bonds in a Bi–Sb mixed crystal (Fig. 3.12) [33] and for the radial breathing mode of single wall carbon nanotubes with different chiralities in ensemble [34]. It can also reveal a weak coherent oscillation, which would otherwise be hidden by other modes. An excellent example was given by Cs–covered Pt surface [35], where the Cs–Pt oscillation dominates the SHG signal pumped with a single pulse. By tuning the repetition rate of the pulse train slightly higher than the Cs-Pt stretching frequency, the Pt surface Rayleigh phonon mode was clearly seen in the FT spectrum (Fig. 3.13).

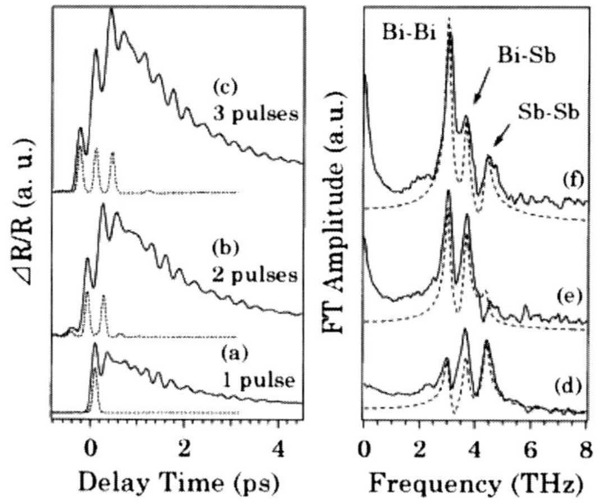

Fig. 3.12. Selective enhancement of coherent A_{1g} phonon of Bi–Bi vibration in $Bi_{0.31}Sb_{0.69}$ mixed crystal at room temperature. (**a**–**c**) reflectivity change in time domain with single, double and triple pulses. *Dotted curves* represent SHG profile of the pump pulses. (**d**–**f**) corresponding FT spectra. From [33]

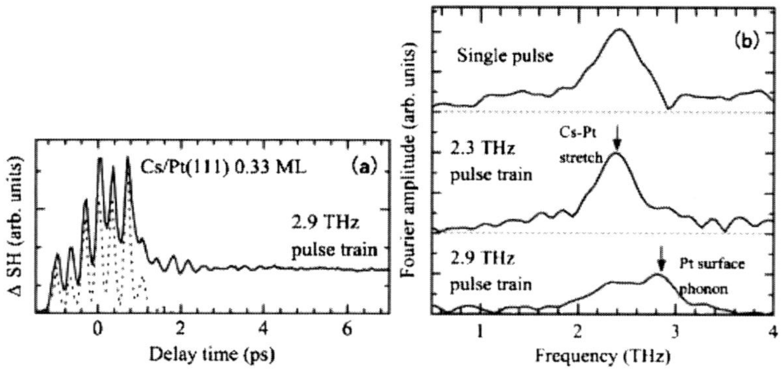

Fig. 3.13. (**a**) Transient SHG signal (*solid curve*) obtained from 0.33 ML Cs on Pt(111) for excitation with 2.9-THz repetition rate pulse-train (*dashed curve*). (**b**) FT spectra of the SHG signal from the same surface obtained with a single pump pulse (*top trace*) and with pulse trains at different repetition rates (*center and bottom traces*). From [35,36]

If the dephasing time of the coherent phonons depend critically on the carrier density, photo-injection of carriers with the second pump pulse can annihilate them partially or completely, depending on its fluence but not on its relative timing. Such "incoherent" control was demonstrated for the LO phonons of GaAs [37].

3.2.2 Optical Control Near the Lindemann Limit

Coherent control experiments were also performed under intense photoexcitation near the Lindemann stability limit [38–41], above which the long-range order of the crystal would be lost.

When irradiated with mJ/ cm^2 pulses, the amplitude of the coherent A_{1g} phonons in Bi was canceled by the second pump pulse at $\Delta\tau = 1.5T$, but was hardly enhanced by that at $\Delta\tau = T$, indicating that the pump fluence was in the "nonlinear regime" [38]. Similar double-pump experiments were performed on Bi to investigate the origin of the transient frequency shift of the coherent A_{1g} phonons [39] and the ultrafast energy transfer from electronic to lattice subsystems [41]. Murray and coworkers demonstrated that the frequency was not dependent on the amplitude of the partially canceled oscillation, but was determined by the pump fluence [39]. The results led to a conclusion that the electronic softening, not the lattice anharmonicity, dominated the redshift of the A_{1g} frequency. However, a later double-pump experiment with an extended range of $\Delta\tau$ by Misochko and coworkers [40] revealed both correlation and anticorrelation between the initial amplitude and the frequency, indicating that the lattice anharmonicity cannot be ignored.

A multiple-pump experiment on Te by Mazur and coworkers revealed that the reflectivity oscillation was enhanced to maximum or canceled completely when $\Delta\tau$ was considerably shorter than nT or $(n+1/2)T$ [32]. In other words, the nuclear vibrations do not stop at their classical turning point, in contrast to the weak excitation case. This departure from a classical harmonic motion is the manifestation of a time-dependent driving force, whose physical origin

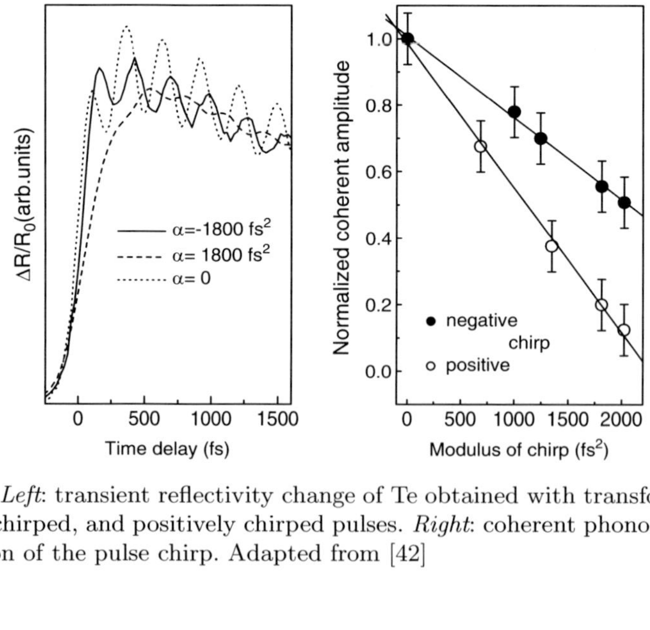

Fig. 3.14. *Left*: transient reflectivity change of Te obtained with transform limited, negatively chirped, and positively chirped pulses. *Right*: coherent phonon amplitude as a function of the pulse chirp. Adapted from [42]

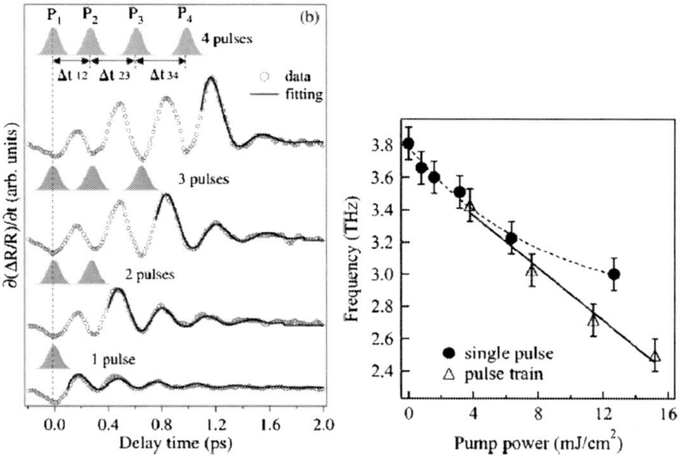

Fig. 3.15. *Left*: reflectivity change of GeTe showing the coherent A_{1g} phonon under repetitive excitation with a pulse train. The intervals between the pump pulses, $\Delta t_{12}, \Delta t_{23}$, and Δt_{34} are 290, 320, and 345 fs, respectively. *Right*: pump power dependence of the frequency of the A_{1g} mode for excitation with a single pulse and the pulse train. From [43]

would possibly involve many body effects, such as a deformation coupling that changes with lattice configuration.

Reversing the sign of the chirp of a single pump pulse did not affect the amplitude of the coherent phonons in Te and Bi at low excitation strength, but only brought a small asymmetry in the dephasing time [42]. In contrast, at high excitation strength there was a clear dependence on the chirp-sign in both Bi and Te; the oscillation amplitude decreased with increasing chirp with a rate that depends on the chirp sign (Fig. 3.14). Even more striking was that the oscillation amplitude appeared to be larger for a slightly negatively chirped pulse than for the transform limited one.

In ferroelectric GeTe, which undergoes the structural phase transition at 657 K, the coherent A_{1g} mode (soft mode) was excited to a large amplitude using optical pulse trains [43]. Under single-pulse excitation, the frequency of the soft mode red-shifted and eventually saturated at high pump powers. By contrast, under a sequence of pump pulses, a large red-shift was observed without saturation (Fig. 3.15). In the latter situation, the coherent phonon was strongly damped, which indicated that the lattice was driven to a precursor state of the phase transition.

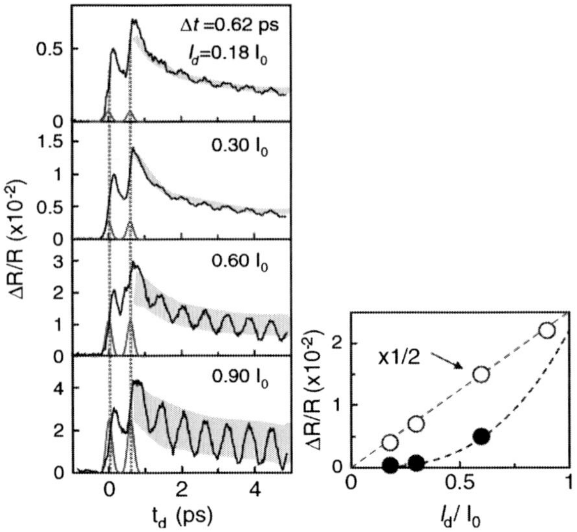

Fig. 3.16. *Left*: transient reflectivity changes of TTF-CA with double pulse excitation at $\Delta\tau = 0.62$ ps ($=T$) for various pump fluences, indicating the super-linear enhancement of the coherent phonons with increasing excitation density. The pump pulses are indicated by *grey lines*. I_d indicates the excitation density of each pulse in the pulse pair in unit of $I_0 = 1 \times 10^{16}$ photons/ cm^2. *Right*: Maximum $\Delta R/R$ (*open circles*) and the oscillation amplitude (*closed circles*) as a function of I_d/I_0. From [44]

3.2.3 Optical Control in Strongly Correlated Systems

Double pump experiments on an organic charge transfer complex TTF-CA by Iwai and coworkers demonstrated a new class of "coherent" control on a strongly correlated electron–lattice system [44]. While the amplitude of the coherent oscillation increased linearly with pump fluence for single pump experiments, the amplitude in the double pump experiments with a fixed pulse interval $\Delta\tau = T$ exhibited a strongly super-linear fluence dependence (Fig. 3.16). The striking difference between the single- and double-pulse results indicated a cooperative nature of the photo-induced neutral-ionic transition.

3.3 Concluding Remarks

In this chapter, we have reviewed the experimental researches on the coherent optical phonons detected by novel nonoptical techniques, as well as on their optical control using shaped pulses. TRXRD can separate atomic displacements from collective motions of charges, while TR-THz and TRPE spectroscopies can visualize the ultrafast development of strong interaction between phonons and electrons in the real time. These new detection techniques are contributing to attract a renewed, broad interest on coherent optical phonons, and have opened a novel direction for the research, especially in highly correlated systems.

Control of the lattice vibrations in crystals has so far been achieved only through classical interference. Optical control in solids is far more complicated than in atoms and molecules, especially because of the strong interaction between the phononic and electronic subsystems. Nevertheless, we expect that the rapidly developing pulse-shaping techniques will further stimulate pioneering studies on optical control of coherent phonons.

Acknowledgements

The authors thank the following people for their valuable contributions to this and the previous chapters: Hrvoje Petek, Uwe Bovensiepen, Alexey Melnikov, Ken Onda, Kazuya Watanabe, Muneaki Hase, Kazutaka G. Nakamura, and Eeuwe Zijlstra.

References

1. K. Sokolowski-Tinten, D. von der Linde, J. Phys. Condens. Mater. **16**, R1517 (2004)
2. K. Sokolowski-Tinten, C. Blome, J. Blums, A. Cavalleri, C. Dietrich, A. Tarasevitch, I. Uschmann, E. Förster, M. Kammler, M. Horn von Hoegen, D. von der Linde, Nature **422**(6929), 287 (2003)
3. D.A. Reis, A.M. Lindenberg, in *Light Scattering in Solids IX*, ed. by M. Cardona, R. Merlin (Springer, Berlin, 2007), page 371
4. K.G. Nakamura, S. Ishii, S. Ishitsu, M. Shiokawa, H. Takahashi, K. Dharmalingam, J. Irisawa, Y. Hironaka, K. Ishioka, M. Kitajima, Appl. Phys. Lett. **93**, 061905 (2008)
5. A.L. Cavalieri, D.M. Fritz, S.H. Lee, P.H. Bucksbaum, D.A. Reis, J. Rudati, D.M. Mills, P.H. Fuoss, G.B. Stephenson, C.C. Kao, D.P. Siddons, D.P. Lowney, A.G. MacPhee, D. Weinstein, R.W. Falcone, R. Pahl, J. Als-Nielsen, C. Blome, S. Dusterer, R. Ischebeck, H. Schlarb, H. Schulte-Schrepping, T. Th, J. Schneider, O. Hignette, F. Sette, K. Sokolowski-Tinten, H.N. Chapman, R.W. Lee, T.N. Hansen, O. Synnergren, J. Larsson, S. Techert, J. Sheppard, J.S. Wark, M. Bergh, C. Caleman, G. Huldt, D. van der Spoel, N. Timneanu, J. Hajdu, R.A. Akre, E. Bong, P. Emma, P. Krejcik, J. Arthur, S. Brennan, K.J. Gaffney, A.M. Lindenberg, K. Luening, J.B. Hastings, Phys. Rev. Lett. **94**(11), 114801 (2005)
6. O.V. Misochko, M. Hase, K. Ishioka, M. Kitajima, Phys. Rev. Lett. **92**(19), 197401 (2004)
7. D.M. Fritz, D.A. Reis, B. Adams, R.A. Akre, J. Arthur, C. Blome, P.H. Bucksbaum, A.L. Cavalieri, S. Engemann, S. Fahy, R.W. Falcone, P.H. Fuoss, K.J. Gaffney, M.J. George, J. Hajdu, M.P. Hertlein, P.B. Hillyard, M. Horn von Hoegen, M. Kammler, J. Kaspar, R. Kienberger, P. Krejcik, S.H. Lee, A.M.

Lindenberg, B. McFarland, D. Meyer, T. Montagne, É.D. Murray, A.J. Nelson, M. Nicoul, R. Pahl, J. Rudati, H. Schlarb, D.P. Siddons, K. Sokolowski-Tinten, T. Tschentscher, D. von der Linde, J.B. Hastings, Science **315**, 633 (2007)
8. S.L. Johnson, P. Beaud, C.J. Milne, F.S. Krasniqi, E.S. Zijlstra, M.E. Garcia, M. Kaiser, D. Grolimund, R. Abela, G. Ingold, Phys. Rev. Lett. **100**(15), 155501 (2008)
9. S.L. Johnson, P. Beaud, E. Vorobeva, C.J. Milne, E.D. Murray, S. Fahy, G. Ingold, Phys. Rev. Lett. **102**(17), 175503 (2009)
10. A. Cavalleri, S. Wall, C. Simpson, E. Statz, D.W. Ward, K.A. Nelson, M. Rini, R.W. Schoenlein, Nature **442**(7103), 664 (2006)
11. K. Ishioka, M. Kitajima, J. Irisawa, Y. Hironaka, K. Ushida, K.G. Nakamura, Jpn. J. Appl. Phys. **45**(12), 9111 (2006)
12. E. Seres, C. Spielmann, Appl. Phys. Lett. **91**, 121919 (2007)
13. M.C. Beard, G.M. Turner, C.A. Schmuttenmaer, Phys. Rev. B **62**(23), 15764 (2000)
14. R. Huber, A. Brodschelm, F. Tauser, A. Leitenstorfer, Appl. Phys. Lett. **76**(22), 3191 (2000)
15. C. Kübler, R. Huber, S. Tubel, A. Leitenstorfer, Appl. Phys. Let. **85**(16), 3360 (2004)
16. R. Huber, C. Kübler, S. Tübel, A. Leitenstorfer, Q.T. Vu, H. Huang, F. Köhler, M.-C. Amann, Phys. Rev. Lett. **94**(02), 027401 (2005)
17. A. Cavalleri, C. Tóth, C.W. Siders, J.A. Squier, F. Ráksi, P. Forget, J.C. Kieffer, Phys. Rev. Lett. **87**(23), 237401 (2001)
18. A. Cavalleri, T. Dekorsy, H.H.W. Chong, J.C. Kieffer, R.W. Schoenlein, Phys. Rev. B **70**(16), 161102 (2004)
19. C. Kübler, H. Ehrke, R. Huber, R. Lopez, A. Halabica, R.F. Haglund Jr., A. Leitenstorfer, Phys. Rev. Lett. **99**(11), 116401 (2007)
20. P.A. Loukakos, M. Lisowski, G. Bihlmayer, S. Blugel, M. Wolf, U. Bovensiepen, Phys. Rev. Lett. **98**(9), 097401 (2007)
21. L. Perfetti, P.A. Loukakos, M. Lisowski, U. Bovensiepen, H. Berger, S. Biermann, P.S. Cornaglia, A. Georges, M. Wolf, Phys. Rev. Lett. **97**(6), 067402 (2006)
22. F. Schmitt, P.S. Kirchmann, U. Bovensiepen, R.G. Moore, L. Rettig, M. Krenz, J.H. Chu, N. Ru, L. Perfetti, D.H. Lu, M. Wolf, I.R. Fisher, Z.X. Shen, Science **321**(5896), 1649 (2008)
23. U. Bovensiepen, J. Phys. Condens. Matter **19**, 235417 (2007)
24. A.M. Weiner, D.E. Leaird, G.P. Wiederrecht, K.A. Nelson, Science **247**(4948), 1317 (1990)
25. M.W. Wefers, H. Kawashima, K.A. Nelson, J. Phys. Chem. Solid **57**(10), 1425 (1996)
26. T. Dekorsy, W. Kutt, T. Pfeifer, H. Kurz, Europhys. Lett. **23**(3), 223 (1993)
27. M. Hase, K. Mizoguchi, H. Harima, S. Nakashima, M. Tani, K. Sakai, M. Hangyo, Appl. Phys. Lett. **69**(17), 2474 (1996)
28. K.G. Lee, D.S. Kim, K.J. Yee, H.S. Lee, Phys. Rev. B **74**, 113201 (2006)
29. A. Bartels, T. Dekorsy, H. Kurz, K. Köhler, Appl. Phys. Lett. **72**(22), 2844 (1998)
30. U. Ozgur, C.-W. Lee, H.O. Everitt, Phys. Rev. Lett. **86**(24), 5604 (2001)
31. Y. Ogawa, F. Minami, Phys. Rev. B **75**, 73302 (2007)
32. C.A.D. Roeser, M. Kandyla, A. Mendioroz, E. Mazur, Phys. Rev. B **70**, 212302 (2004)

33. M. Hase, T. Itano, K. Mizoguchi, S. Nakashima, Jpn. J. Appl. Phys. **37**(3A), L281 (1998)
34. J.H. Kim, K.J. Han, N.J. Kim, K.J. Yee, Y.S. Lim, G.D. Sanders, C.J. Stanton, L.G. Booshehri, E.H. Haroz, J. Kono, Phys. Rev. Lett. **102**(3), 037402 (2009)
35. K. Watanabe, N. Takagi, Y. Matsumoto, Phys. Chem. Chem. Phys. **7**(14), 2697 (2005)
36. Y. Matsumoto, K. Watanabe, Chem. Rev. **106**(10), 4234 (2006)
37. K.J. Han, J.H. Kim, D.W. Jang, K.J. Yee, X. Liu, J.K. Furdyna, Y.S. Lim, J. Korean Phys. Soc. **50**(3), 781 (2007)
38. M.F. DeCamp, D.A. Reis, P.H. Bucksbaum, R. Merlin, Phys. Rev. B **64**, 092301 (2001)
39. E.D. Murray, D.M. Fritz, J.K. Wahlstrand, S. Fahy, D.A. Reis, Phys. Rev. B **72**, 060301(R) (2005)
40. O.V. Misochko, R. Lu, M. Hase, M. Kitajima, JETP **104**(2), 245–253 (2007)
41. A.Q. Wu, X. Xu, Appl. Phys. Lett. **90**(25), 251111 (2007)
42. O.V. Misochko, T. Dekorsy, S.V. Andreev, V.O. Kompanets, Y.A. Matveets, A.G. Stepanov, S.V. Chekalin, Appl. Phys. Lett. **90**(7), 071901 (2007)
43. M. Hase, M. Kitajima, S. Nakashima, K. Mizoguchi, Appl. Phys. Lett. **83**(24), 4921 (2003)
44. S. Iwai, Y. Ishige, S. Tanaka, Y. Okimoto, Y. Tokura, H. Okamoto, Phys. Rev. Lett. **96**(5), 057403 (2006)

4

Heterodyne Interferometry Using High-Order Harmonic Generation in Mixed Gases

Tsuneto Kanai, Eiji J. Takahashi, and Yasuo Nabekawa,
and Katsumi Midorikawa

Summary. We describe a new and simple method to measure harmonic phases by using the interference of harmonics in mixed gases both experimentally and theoretically. First, we develop a fundamental theory of HHG in mixed gases by combining the propagation equation of high-order harmonics and the Lewenstein model, which describes the single atom (molecule) response. The harmonic phase can be obtained by comparing the harmonic spectra from the two gases and their mixture. Then, we present the first experimental evidence of the destructive and constructive interference in a mixed gas of He and Ne, which prove the validity of the method and facilitate the coherent control of HHG. Finally, we applied this method to the attosecond physics. By measuring the relative phase of high-order harmonics generated in each molecule, attosecond dynamics of nuclear wavepackets in H2 and D2 molecules can be probed. The measured relative phase as well as intensity ratio are successfully reproduced by applying the Feynman's path integral method fully to the dynamics of the nuclei and electrons in the molecules.

4.1 Introduction

High-order harmonic generation (HHG) in atoms and/or molecules using ultrashort lasers has been a subject of intense studies for its applications as a coherent attosecond radiation source in the extreme ultraviolet (XUV) and soft X-ray regions [1–7]. The physics of HHG, including the amplitude and the phase of harmonics, are well understood by the three-step model (Fig. 4.1) [8]. First, a part of the bound-state electron wavefunction tunnels through the potential barrier modified by the intense laser field and appears in the continuum (step 1). The freed electron wavepacket is then driven by the laser field, and after the field reverses its direction, it has a probability of returning to the parent ion with a quantum mechanical phase, which depends on the excursion time τ_j of the trajectory j and the ionization potential I_{p}^i of the atom or the molecule i (step 2). The harmonic photon is emitted by the coherent oscillation between returning electron wavepacket and the

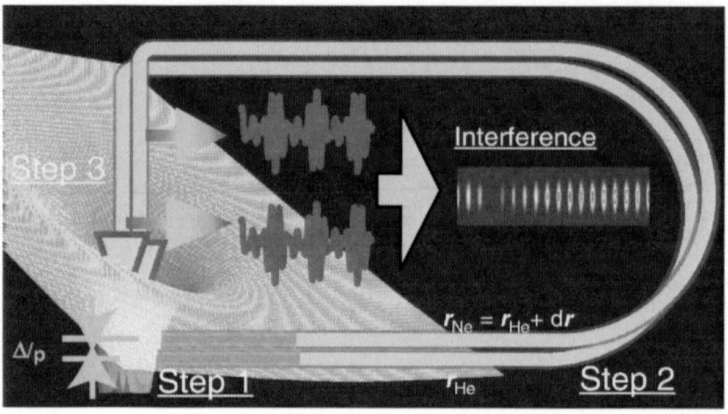

Fig. 4.1. An illustration of two quantum paths of electrons that contribute the HHG in a mixed gas. The variation on quantum path $\delta\mathbf{r}$ generates the phase difference of harmonics. Notice that ΔI_{p} expresses the difference in energy between the two potentials

bound-state electron wavefunction at time t (step 3). The harmonics generated at the sites where the event happened by chance, interfering with each other, propagate through the gas medium under the influence of dispersion and absorption of the medium.

Measuring the phases of high harmonics has been a crucial challenge in attosecond science, as it proves that the harmonic pulses with broad spectra are indeed temporarily confined in attosecond regions and harmonic phases contain crucial information on the instantaneous electronic structures of the nonlinear media. To measure phases of harmonics, however, state-of-the-art methods such as RABITT (reconstruction of attosecond beating by interference of two-photon transitions [9]), PANTHER (the photoelectron analysis with nonresonant two-photon-ionization for harmonic electric-field reconstruction) [4], FROG (frequency-resolved optical gating) [10], and FROG-CRAB(Complete Reconstruction of Attosecond Bursts) were needed, and in fact, only several groups succeeded in obtaining this remaining grail.

In the present chapter, we describe a new and simple method to measure harmonic phases by using the interference of harmonics in mixed gases [11,12]. In Sect. 4.2, we develop a fundamental theory of HHG in mixed gases by combining the propagation equation of high-order harmonics and the generalized Lewenstein model [8], which describes the single atom (molecule) response. The harmonic phase can be obtained by comparing the harmonic spectra from the two gases and their mixture. In Sect. 4.3, we present the first experimental evidence of the destructive interference (DI) and the constructive interference (CI) in a mixed gas of He and Ne, which prove the validity of the method and facilitate the coherent control of HHG [13]. In Sect. 4.4, we applied this method to the attosecond physics. By measuring the relative phase of high-order harmonics generated in each molecule, attosecond dynamics of nuclear wavepackets in H_2 and D_2 molecules can be probed. The measured relative phase as well as intensity ratio are successfully reproduced by applying the Feynman's path integral method [14, 15] fully to the dynamics of the nuclei and electrons in the molecules [11].

4.2 Theory of HHG in Mixed Gases and Heterodyne Interferometry for Detection of Ultrafast Molecular Dynamics

Consider the power spectra of the harmonics from gas 1, 2, and mixture of them, $\mathcal{P}_1(\omega)$, $\mathcal{P}_2(\omega)$, and $\mathcal{P}_{1+2}(\omega)$. As can be seen in the simplest case of two point emitters at the site 1 and 2, $|d_1(\omega)+d_2(\omega)|^2 = |d_1(\omega)|^2 + |d_2(\omega)|^2 + 2\mathrm{Re}[d_1(\omega)d_2^*(\omega)]$, where $d_i(\omega)$ is the harmonic amplitude at the site i, the effects of the interference between the two HHG processes (i.e., the counterpart of $2\mathrm{Re}[d_1(\omega)d_2^*(\omega)]$) should appear in the difference between $\mathcal{P}_{1+2}(\omega)$ and $\mathcal{P}_1(\omega)+\mathcal{P}_2(\omega)$. By integrating the propagation equation of harmonics and the Schrödinger equation of the active electrons, the spectrum $\mathcal{P}_{\Sigma i}(\omega)$ from the mixed gas composed of gases i's at the qth harmonics ($\omega = q\omega_0 \equiv \omega_q$ with the photon energy of the fundamental field ω_0) emitted on axis can be derived as

$$
\begin{aligned}
\mathcal{P}_{1+2}(\omega_q) &\propto \mathcal{I}_{\mathrm{gen}}(\omega_q) \times \mathcal{I}_{\mathrm{prop}}(\omega_q) \\
\mathcal{I}_{\mathrm{gen}}(\omega_q) &:= (\rho_{\mathrm{tot}} L)^2 \left|\langle d(\omega_q)\rangle\right|^2 \\
\mathcal{I}_{\mathrm{prop}}(\omega_q) &:= \frac{(2L_{\mathrm{a}})^{-2} + \pi^2 L_{\mathrm{c}}^{-2}}{1 + e^{-L_{\mathrm{a}}^{-1}} - 2\cos(\pi L_{\mathrm{c}}^{-1}) e^{-(2L_{\mathrm{a}})^{-1}}},
\end{aligned}
\tag{4.1}
$$

where $\rho_{\mathrm{tot}} := \sum_i \rho_i$, with ρ_i being the density of the gas i, L is the medium length, $\langle\cdots\rangle$ denotes the density weighted average, for example, $\langle d(\omega_q)\rangle = \sum_i r_i d_i(\omega_q)$, with $r_i := \rho_i/\rho_{\mathrm{tot}}$. $L_{\mathrm{a}}(q) := \left[2\rho_{\mathrm{tot}} L \lambda_0 r_e \langle f_2\rangle / q\right]^{-1}$ is the dimensionless absorption length, where $\lambda_0 := 2\pi c/\omega_0$, r_e is the classical electron radius, and f_2 is the imaginary part of the atomic scattering factor. $L_{\mathrm{c}}(q) := \pi/(\Delta k_{\mathrm{tot}}(\omega_q) L)$ is the dimensionless coherence length with $\Delta k_{\mathrm{tot}} := k_q - q k_0$, where k_q and k_0 are the wave vectors of the qth harmonics and of the fundamental field, respectively [16].

The first term in (4.1), $\mathcal{I}_{\mathrm{gen}}(\omega_q)$, expresses the generation of harmonics by integrated single atom responses interfering with each other. In the two-gas case, it can be simplified as

$$
\begin{aligned}
\mathcal{I}_{\mathrm{gen}}(\omega_q) = {} & \rho_1^2 |d_1(\omega_q)|^2 + \rho_2^2 |d_2(\omega_q)|^2 \\
& + 2\rho_1\rho_2 |d_1(\omega_q)||d_2(\omega_q)| \cos\Delta\varphi(\omega_q),
\end{aligned}
\tag{4.2}
$$

where $\Delta\varphi(\omega_q) := \varphi_2(\omega_q) - \varphi_1(\omega_q)$ is the relative phase of the harmonics from the two gases 1 and 2. Straightforwardly, the condition for the CI ($\mu = 0$) and the DI ($\mu = 1$) can be derived as $\Delta\varphi(\omega_q) = (2n + \delta_\mu^1)\pi$ $(n \in \mathbb{Z})$, with the Kronecker delta δ_ν^μ. These interferences are the most fundamental phenomena of HHG in mixed gases; they always appear where the "two" spectra overlap. It should be noted that when the condition for DI is satisfied for the qth harmonics, we can completely suppress HHG by merely setting the density ratio

$$r_2 = \frac{1}{1 + [d_2(\omega_q)/d_1(\omega_q)]}$$

regardless of the propagation condition.

Here we restrict our attention to the simplest and ordinary case in which $d_i(\omega_q)$ can be well described by the nonlinear response to the fundamental pulse only. As can be confirmed in [11], this is in the case of a He–Ne mixed gas. If the density of generated harmonic photons in the interaction region is not low enough compared to that of the fundamental photons, $d_i(\omega_q)$ should strongly be influenced by the generated harmonic fields [17–19]. Using the saddle point method $\delta S_i(\boldsymbol{p}, t, \tau) = 0$, where $S_i(\boldsymbol{p}, t, \tau)$ is the action of the active electron in the atom i, the stationary phase S_i^{s} is found to be a linear function of I_{p}^i and the ponderomotive potential U_{p}: $S_i^{\mathrm{s}} = \tau^{\mathrm{s}} I_{\mathrm{p}}^i + [\tau^{\mathrm{s}} - 2(1 - \cos\tau^{\mathrm{s}})/\tau^{\mathrm{s}} - C(\tau^{\mathrm{s}})\cos(2t^{\mathrm{s}} - \tau^{\mathrm{s}})]U_{\mathrm{p}}$, where $C(\tau) := \sin(\tau) - 4\sin^2(\tau/2)/\tau$ [8].[1] Considering $U_{\mathrm{p}} \propto I_{\mathrm{IR}}$, atomic dipole phases linearly depend on the laser intensity I_{IR} and the dependence attracted much attention for the potential to measure the harmonic phases [10, 20–22], although the underlying parameter of laser intensity cannot be controlled precisely. We offer an alternative approach by using the other dependence, that is, the dependence on I_{p}, which is the intrinsic quantity determined by Nature exactly. By preparing two gases with different I_{p}'s, we can measure the phase of harmonics in a robust way. In fact, when the short trajectory is selected as in our case [7], the relative phase $\Delta\varphi(\omega_q)$ takes a simple form,[2]

$$\Delta\varphi(\omega_q) \approx \tau_1(\omega_q)\Delta I_{\mathrm{p}}. \tag{4.3}$$

The first order term in ΔI_{p} comes from the difference of the potential energy and the higher order terms should be included when $\Delta I_{\mathrm{p}}/U_{\mathrm{p}}$ is not small enough. The phases, which the freed electrons accumulate during their different quantum paths, are transferred to the harmonics through the coherent process of HHG and lead to the interferences (Fig. 4.1).

The dimensionless second term in (4.1), $\mathcal{I}_{\mathrm{prop}}(\omega_q)$, expresses the role of propagation in mixed gases. As in the case of HHG in pure gases [23], the optimization conditions $1/3 > L_{\mathrm{a}}$ and $L_{\mathrm{c}} > 5L_{\mathrm{a}}$ ensure that the macroscopic response is more than half of the maximum response. The asymptotic value in $L \to \infty$ and $L_{\mathrm{c}} \to \infty$ is independent of ρ_{tot} and is a monotonically increasing

[1] When $I_{\mathrm{p}} \neq 0$, τ^{s} takes complex values and accordingly, we define the excursion time $\tau_1 := \mathrm{Re}(\tau_1^s)$. Notice that the classical simulations do not provide correct τ's [20].

[2] For short trajectory, the remaining correction term that is proportional to U_{p} can be ignored.

function of $|\langle d(\omega_q)\rangle/\langle\sigma(\omega_q)\rangle|^2$, where $\langle\sigma(\omega_q)\rangle \propto L_{\mathrm{a}}^{-1}$ is the expectation value of the ionization cross section. Optimizing HHG requires to simultaneously maximize $|\langle d(\omega_q)\rangle|/\langle\sigma(\omega_q)\rangle$ and to fulfill the above conditions. Considering the interference effect on $|\langle d(\omega_q)\rangle|$, the optimization conditions for HHG in mixed gases are now given by

$$\begin{aligned} \Delta\varphi(\omega_q) &\approx 2n\pi \quad (n \in \mathbb{Z}), \\ 1/3 &> L_{\mathrm{a}}, \\ L_{\mathrm{c}} &> 5L_{\mathrm{a}}. \end{aligned} \tag{4.4}$$

By using this condition, we demonstrate enhancing HHG in a mixed gas of He and Ne in [11].

Finally from (4.1), the relative phase can be written as

$$\Delta\varphi(\omega_q) = \cos^{-1}\left[\frac{\mathcal{P}^{1+2}/\mathcal{I}_{\mathrm{prop}}^{1+2} - \mathcal{P}^1/\mathcal{I}_{\mathrm{prop}}^1 - \mathcal{P}^2/\mathcal{I}_{\mathrm{prop}}^2}{2[\mathcal{P}^1\mathcal{P}^2/(\mathcal{I}_{\mathrm{prop}}^1\mathcal{I}_{\mathrm{prop}}^2)]^{1/2}}\right]. \tag{4.5}$$

Therefore, if the condition of the nonlinear media is controlled, the relative phase $\Delta\varphi(\omega_q)$ can be determined by observing harmonic spectra $\mathcal{P}^1$, $\mathcal{P}^2$, and $\mathcal{P}^{1+2}$.

4.3 Destructive Interference During HHG in Mixed Gases

In this section, we present the first experimental evidence of the destructive interference (DI) and the constructive interference (CI) in a mixed gas of He and Ne, which prove the validity of the method. The observed interference modulation is, as discussed in Sect. 4.2, attributed to the difference between the phases of the intrinsically chirped harmonic pulses from He and Ne, which leads to the novel method for broadband measurement of the harmonic phases and for observing the underlying attosecond electron dynamics.

4.3.1 Experimental

We performed an experiment with a Ti:sapphire laser having the center wavelength $\lambda_0 \approx 800$ nm, the maximum energy ≈200 mJ, and the pulse duration ≈30 fs. The fundamental pulse was loosely focused with an $R = 10$ m concave mirror into an interaction gas cell filled with a pure and/or a mixed gas, and the macroscopic effects $\mathcal{I}_{\mathrm{p}}^l$ are precisely controlled by stabilizing the condition of the gases (Fig. 4.2). The generated harmonics are spectrally resolved with a flat-field grating and detected by a microchannel plate and a CCD camera. See [7] for the details of our setup. The intensity of the fundamental pulses was estimated to be $I_{\mathrm{IR}} \approx 4 \times 10^{14}$ W cm^{-2}, which is below the saturation

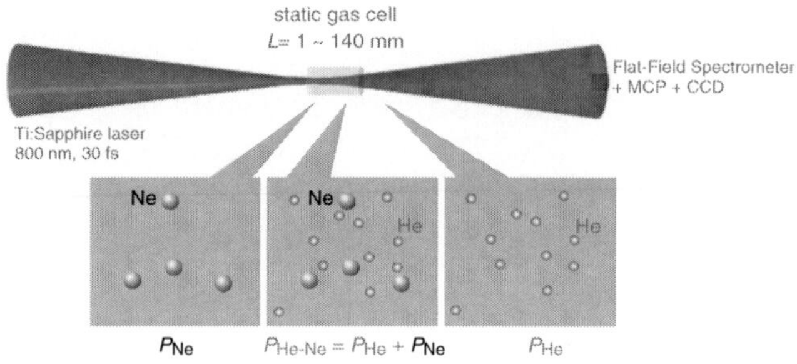

Fig. 4.2. Experimental setup

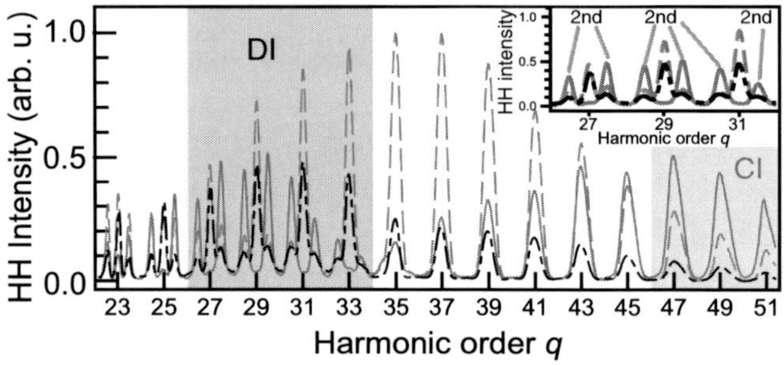

Fig. 4.3. The harmonic spectra from a 13.5 Torr He gas (*dot-dashed curve*), a 1.5 Torr Ne gas (*dashed curve*), and the mixture of them (*thick curve*). Inset shows the enlarged illustration of the spectra, where DI takes place

of ionization of atoms, ensuring that harmonics are generated mostly at the peak of the laser pulse. The medium length was set to be $L \approx 13\,\mathrm{cm}$.

Figure 4.3 shows the typical harmonic spectra in He ($p_{\mathrm{He}} = 13.5\,\mathrm{Torr}$), Ne ($p_{\mathrm{Ne}} = 1.5\,\mathrm{Torr}$), and the mixture of them ($r_{\mathrm{Ne}} = 10\%$, $p_{\mathrm{tot}} = 15.0\,\mathrm{Torr}$). The second-order diffraction rays of the 45th–71st harmonics appeared at the side of the 23rd–35th harmonics. As expected, an interference modulation is observed in the spectra from the mixed gas: The harmonics around 49th are enhanced and the harmonics around 29th are suppressed. It should be emphasized that $\mathcal{P}_{\mathrm{He+Ne}}(\omega_{29})$ is completely suppressed compared to $\mathcal{P}_{\mathrm{He}}(\omega_{29}) + \mathcal{P}_{\mathrm{Ne}}(\omega_{29})$. Physically, this spectral modulation should correspond to the interference between the harmonics from He and Ne. In fact, with the help of the (4.1), the experimental results can be translated to the interference modulation characterized by relative phase $\Delta\varphi(\omega_q)$ of harmonics. Figure 4.4 shows the relative phase $\Delta\varphi(\omega_q)$ as a function of harmonic order q. DI and CI take place when $\Delta\varphi(\omega_q) \approx \pi$ and 0, respectively, and within the experimental errors due to the laser fluctuation and the ground state depletion, we see satisfactory agreement between the experimental results and the theoretical calculations, which ensures the validity of our interpretation and model.

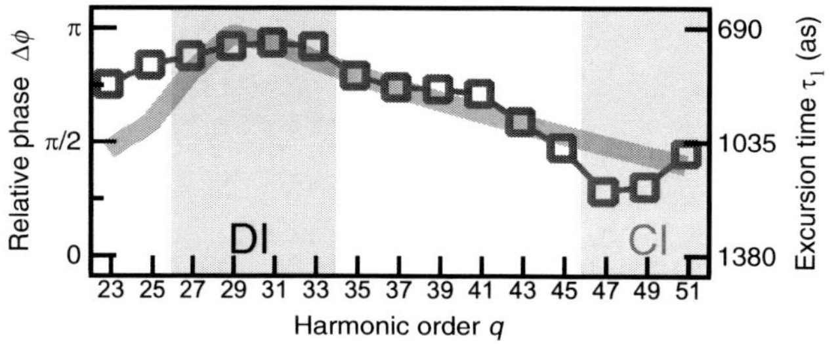

Fig. 4.4. The experimentally determined relative phases of harmonics and the corresponding excursion time τ_1 as a function of harmonic order (*squares*). The error bars are about the sizes of the symbols for the experimental data. Theoretically obtained curve are also shown by the *gray* curve

One of the important consequences of these results can be revealed when one regards the observed interference as an inverse problem. In fact, from (4.3), observing the relative phase corresponds to the broadband measurement of the excursion time τ_1 and thus the estimation of the individual harmonic phase $\varphi_i(\omega_q)$ [24]. Measuring attosecond excursion times offer a crucial basis for the methods of probing the ultrafast dynamics of molecules by using electron wavepacket [26–28, 55]. Directly from the DI condition, we obtain $\tau_1(\omega_{29}) \approx 690$ fs, which is the shortest duration of particle motion observed to date [2, 24]. It should be noted that the attosecond dynamics of electrons in HHG can be realized only in high intensity laser field and at lower order harmonics. Accordingly, the excursion time of $\tau_1(\omega_{19}) = 1.2$ fs by Salières et al. using the lower intensity of $I_{\mathrm{IR}} = 1.35 \times 10^{14}$ W/cm^2 [24] is much longer than our results. Our approach, keeping the advantages of using interference phenomena which can be caused even by weak harmonics and need no assumption of atomic physics in the measurement stage, minimizes the effect of path fluctuations on the process. Because of the extremely short wavelength of harmonics, path fluctuations inevitably give effects even on the fine experiment by Corci et al. [21]. In our case, the "two" pulses, whose relative phase is of importance, propagate through the same path and phase fluctuations are then automatically canceled. The chirped relative phase required for spectral interference, which is easily controlled with *a delay stage* and a nonlinear optics in the IR regions [29], is realized in a robust manner using the HHG process directly.

Another important consequence is the demonstration of tunable bandpass *filtering* using mixed gases. Bandpass optics have been crucial experimental elements in attosecond physics: they are used to compress harmonic pulses [9, 20], to use harmonics as a tool for energy-dependent processes in matters [3], and to generate single attosecond pulses [1, 2]. Usual bandpass optics, however, inevitably reduce the number of the desired harmonic photons during propagating through the dissipating material and their band cannot be controlled easily. From (4.3), the energy where CI takes place is given by

$\omega = \tau_1^{-1}(2n\pi/\Delta I_p)$, which directly indicate the dependence on ΔI_p in addition to the obvious dependence on I_{IR}. By using a pair of atoms or molecules with different ΔI_p, our method can be used for gaseous tunable bandpass *filtering* in the XUV and soft x-ray regions without passing any dissipating optics. While multilayer optics use interferences due to the path differences between the boundaries based on the fine deposition technologies, our novel *filtering* uses interferences due to the phase difference directly made from the underlying ultrafast electron dynamics in HHG.

4.4 Application of the Heterodyne Interferometry to Attosecond Physics

In Sect. 4.3, we discussed the role of ionization potential in harmonic phase I_p [8,11]. As the geometric structures of the atomic or molecular valence orbitals should give additional effects to the harmonic phases [26,55], by measuring the harmonic phases as well as intensities from a pair of atoms and/or molecules with the same I_p such as H_2–D_2 [12], N_2–Ar, and CO_2–Kr [12, 30], the net effects due to the structures of them and their dynamics can be extracted [11] (Fig. 4.5). The main advantage of the present method, which we call heterodyne interferometry, is that it makes use of the interference of harmonics, and faint signals from the dynamics in the atoms and molecules can be observed.

In this section, by applying the heterodyne interferometry to a mixed gas of H_2 and D_2 molecules, we probe attosecond dynamics of nuclear wavepackets in the molecules. We find that not only the single molecule responses but also the propagation effects of harmonics differ between the two isotopes and that to discuss dynamics of molecules in the single molecule responses, the propagation effects need to be excluded from the raw harmonic signals. The measured relative phase as well as intensity ratio are found to be monotonic functions of the harmonic order and are successfully reproduced by applying

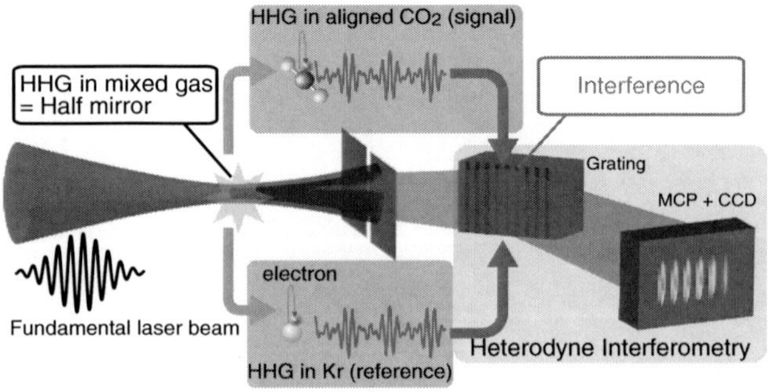

Fig. 4.5. Experimental setup of heterodyne interferometry by using HHG in mixed gases

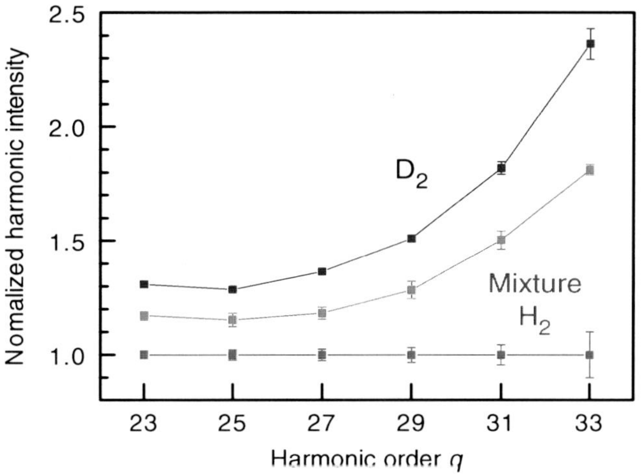

Fig. 4.6. The integrated signals of harmonics from an H_2 gas (10.0 Torr), a D_2 gas (10.0 Torr), and a gas of their mixture (10.0 Torr) measured as a function of the harmonic order q. The signals are normalized to those from the H_2 gas and vertical errors represent SEM (standard error of the mean) for 800 laser shots

the Feynman's path integral method fully to the dynamics of the nuclei and electrons in the molecules.

4.4.1 Results and Discussions

The experiment was performed following the procedure described in Sect. 4.4. Here we used a lens ($f = 2,400$ mm) rather than a focusing mirror ($f = 5{,}000$ mm), and to determine the propagation effect $\mathcal{I}_{\text{prop}}(\omega_q)$, a short gas cell ($L = 15$ mm) was adopted for HHG. The on-target intensity of the fundamental pulses was estimated to be $I_{\text{IR}} \approx 2 \times 10^{14}\,\text{W cm}^{-2}$.

Figure 4.6 shows the integrated signals of harmonics from an H_2 gas (10.0 Torr), a D_2 gas (10.0 Torr), and a mixture of them (10.0 Torr; the mixing ratio = 1:1) measured as a function of the harmonic order q. As the results by Baker et al. [27], the observed harmonic intensity from the D_2 gas is significantly larger than that from the H_2 gas and the information on the relative phase of harmonics is encoded in that from the mixed gas [11].[3] The observed harmonic signals are, however, the integration of the single molecule response and the propagation effects of harmonics and the latter effects must be excluded to discuss the nuclear dynamics in the single molecule responses.

The propagation effects of harmonics, which are due to the two processes of the photoabsorption by each gas and the phase (mis-)matching between the harmonics and the fundamental pulse, can be determined by observing harmonic intensities as functions of the gas pressure [11]. Figure 4.7 shows the

[3] For form factors, see e.g., http://physics.nist.gov/PhysRefData/.

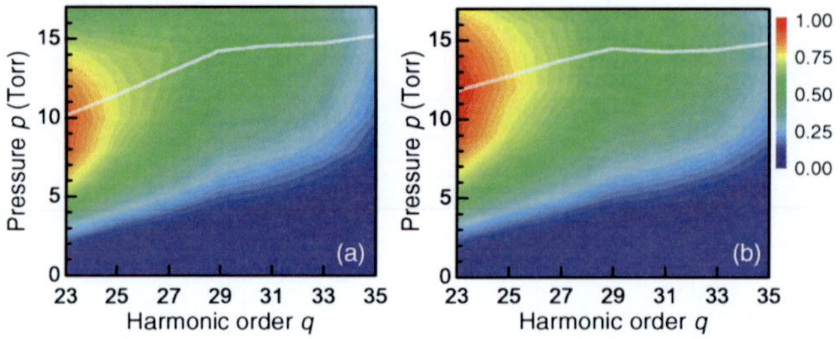

Fig. 4.7. The measured harmonic intensities from (**a**) H_2 and (**b**) D_2 gases as functions of harmonic order q and gas pressure p. The *solid lines* indicate the phase matching pressures, which maximize the harmonic intensities

harmonic intensities from (1) H_2 and (2) D_2 gases as functions of harmonic order q and gas pressure p. A significant difference can be seen between them, that is, the responses in the propagation processes differ between H_2 and D_2 gases. In particular, the difference in the phase matching effects can be seen directly in the ridges in Fig. 4.7, which indicate the phase matching pressures p^i_{PM}; for example, for the 23rd harmonics, $p^{\mathrm{H}_2}_{\mathrm{PM}} \approx 10\,\mathrm{Torr}$ and $p^{\mathrm{D}_2}_{\mathrm{PM}} \approx 12\,\mathrm{Torr}$, which indicate the dispersion of H_2 is 1.2 times as large as that of D_2. Considering the accuracy of the pressure measurement (SEM $\approx 0.02\,\mathrm{Torr}$), this is the first clear observation indicating the differing phase matching effects between the two isotopes. Furthermore, this observation also indicates the other factor of different photoabsorption cross sections indirectly via the Kramers–Kronig relation, and in fact, the photoabsorption cross section of H_2 molecules is significantly larger than that of D_2 molecules in this spectral region [31].[4] As the larger photoabsorption cross section of H_2 molecules by itself can explain the observed smaller harmonic signals from them qualitatively, one should be careful to evaluate the net magnitude of the enhancement.

Disentangling the single molecule responses and the propagation effects of harmonics is crucial especially for measuring the relative phase $\Delta\varphi(q)$ of harmonics. Ignoring the propagation effects of the harmonics leads to inconsistent solutions of imaginary $\Delta\varphi(q)$. For intensity ratio $r(q)$, ignoring the propagation effects leads to significant overestimation of $r(q)$ (Fig. 4.6). Figure 4.8 shows (1) the intensity ratio $r(q)$ in the single molecule responses and (2) the relative phase $\Delta\varphi(q)$ as functions of harmonic order q. As demonstrated by the experiment in [11], the difference in excursion times between successive harmonic orders corresponds to $\approx$100 attoseconds and correspondingly the information on the nuclear dynamics is encoded in the harmonic order (superior axis in Fig. 4.8b). Considering that $r(q)$ and $\Delta\varphi(q)$ are accumulated during the excursion of the electron, it is natural that they are monotonically increasing functions of the excursion time and thus the

[4] It is confirmed directly by measuring the harmonic intensities as functions of the length of the media.

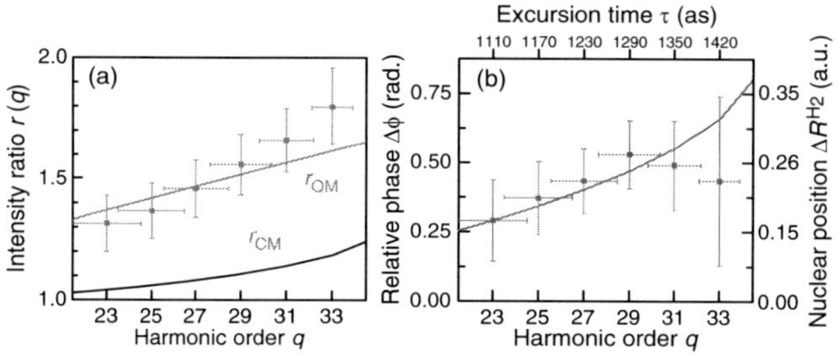

Fig. 4.8. (**a**) The measured (*squares*) and predicted (*thick lines*) intensity ratio $r(q)$ between harmonics from H_2 and that from D_2 molecules as functions of harmonic order q. r_{QM} and r_{CM} are the predictions with and without quantum effect for nuclear dynamics; see text. (**b**) The measured (*squares*) and predicted (*thick line*) relative phases between harmonics from H_2 and those from D_2 molecules as functions of harmonic order q. Measuring the relative phases directly corresponds to observing the nuclear motions, i.e., the nuclear displacement of H_2 ΔR^{H_2} (*right axis*) as a function of excursion time τ (*superior axis*). For (**a**) and (**b**), vertical and horizontal errors represent SEM for 800 laser shots and those from quantum mechanical uncertainty [27]

harmonic order. These results are well reproduced by our model described in the following except for the cutoff region, where the signals are weak and HHG are dominated by the interference between a pair of nearly equal excursion times [32].

4.4.2 Physical Origin of the Interference Signal

The best way to extract physics from the present experiment quantitatively is applying the Feynman's path integral method [14,15] fully to the dynamics of the nuclei and the electron. The action that determines HHG in a molecule i reads

$$S^i_{\mathrm{tot}} = S^i_{\mathrm{free}} + S^i_{\mathrm{e-n}} + S^i_{\mathrm{n}} + S_{\mathrm{rad}} + S_{\mathrm{int}} \,. \tag{4.6}$$

Here $S^i_{\mathrm{free}} + S^i_{\mathrm{e-n}} \equiv S^i_{\mathrm{e}}$ is the action of the active electron used in the Lewenstein's model, which indicates that a few trajectories are responsible for HHG. Among them, the so-called short trajectory, which starts at time t'_{s} and ends at time $t_{\mathrm{s}} = t'_{\mathrm{s}} + \tau_{\mathrm{s}}$, can be selected by using propagation effects.[5] $S^i_{\mathrm{e-n}} \approx \int_0^{\tau_{\mathrm{s}}} \mathrm{d}\tau I_{\mathrm{p}}(R^i(\tau))$ determines the interaction between molecular ion and the electron, where $I_{\mathrm{p}}(R)$ is the ionization potential of the molecules as a function of internuclear distance R^i and was assumed to be constant in [32]. $S^i_{\mathrm{n}} = \int_0^{\tau_{\mathrm{s}}} \mathrm{d}\tau [P^2/(2M^i) - V(R^i(\tau))]$ is the action of the nuclei with the reduced mass M^i, the canonical momentum P, and the potential energy

[5] For complex time, integrand is extended to complex function using analytic continuation.

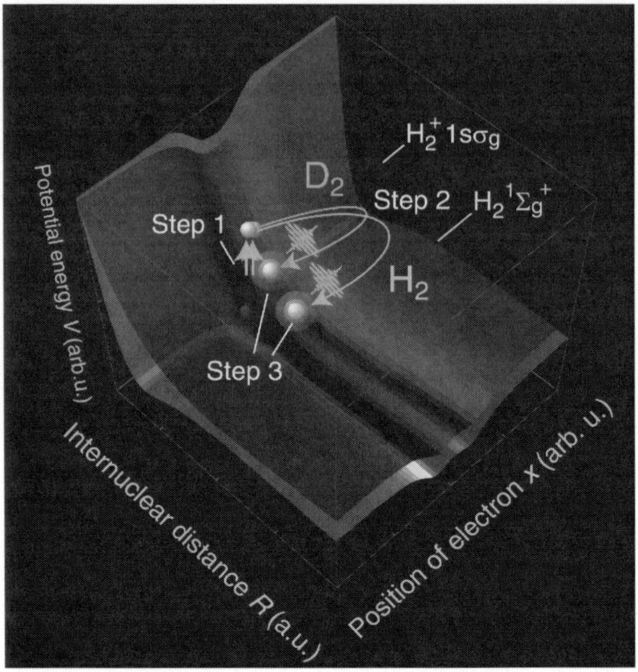

Fig. 4.9. A schematic diagram illustrating the attosecond dynamics of the strongly correlated nuclear and electron wavepackets that lead to HHG in H_2 and/or D_2. When the internuclear distance is large, HHG is suppressed

$V(R^i)$. Finally, $S_{\rm rad}$ and $S_{\rm int}$ are the actions of radiation oscillator and that of the interaction between matter and radiation field, respectively [14, 15].

When a harmonic photon is observed, the relevant quantum path of the active electron and nuclei in the molecules is determined. The nuclei can move after ionization due to their exceptionally light masses (Fig. 4.9a). Lein determines the potential $V(R)$ for their dynamics, which can be approximated by the lowest potential of H_2^+ with genetic algorithm (GA) to fit to the full numerical results [32]. To use GA, however, the solution space must be assumed by hand, and one of the advantages of our method is that such assumptions can be excluded. Considering that the duration $\tau_{\rm s}$ is short enough, the Feynman's propagator $Z^i(\mathcal{R}, t_{\rm s}; R, t'_{\rm s}) := \int \mathcal{D}P\mathcal{D}R\, e^{i(S^i_{\rm n}+S^i_{\rm e-n})} \approx \int \mathcal{D}P\mathcal{D}R\, e^{iS^i_{\rm n}}$, which determines the quantum dynamics of nuclei, is given by [14, 15]

$$Z^i(\mathcal{R}, t_{\rm s}; R, t'_{\rm s}) = \left(\frac{M^i}{2\pi i\tau_{\rm s}}\right)^{\frac{1}{2}} \exp\left[i\left[\frac{M^i(\mathcal{R}-R)^2}{2\tau_{\rm s}} + \frac{f_0\tau(\mathcal{R}+R)}{2} - \frac{f_0^2\tau_{\rm s}^3}{24M^i}\right]\right], \tag{4.7}$$

where $f_0 := -V'(R_0)$ with the equilibrium internuclear distance $R_0 := R(0)$. The initial state can be well approximated by

$$w^i(R,t'_{\rm s}) = N \exp\left[-\frac{(R-R_0)^2}{(\sqrt{2}\delta R^i_0)^2}\right],$$

where $R_0^{\rm H_2} \approx R_0^{\rm D_2} = 7.4\times10^{-2}$ nm $\equiv R_0$ and $\delta R_0^{\rm H_2,D_2} = 8.7\times10^{-3}, 7.5\times10^{-3}$ nm express quantum fluctuations of the initial states. Then the probability density in the recombination is given by

$$|w^i(R,t_{\rm s})|^2 = N\left(1+\frac{\tau_{\rm s}^2}{(\delta R^i_0)^4(M^i)^2}\right)^{-\frac{1}{2}} \exp\left[-\frac{(R-R^i_{\rm cl}(\tau))^2/(\delta R^i_0)^2}{1+\tau_{\rm s}^2/(\delta R^i_0)^4(M^i)^2}\right], \tag{4.8}$$

where $R^i_{\rm cl}(\tau_{\rm s}) := R_0 + f_0\tau_{\rm s}^2/(2M^i) \equiv R_0 + \Delta R^i(\tau_{\rm s})$ corresponds to the classical trajectory of uniformly accelerated nuclear motion. This evolution of $R^i_{\rm cl}(\tau_{\rm s})$ results in the variation of electron dynamics that lead to HHG. The quantum fluctuation is given by $\delta R^i(\tau_{\rm s}) := \delta R^i_0[1+\tau_{\rm s}^2/(\delta R^i_0)^4(M^i)^2]^{1/2}$, which is due to the fluctuation at the initial state and quantum diffusion during the dynamics on the potential curve. This quantum treatment of the nuclear dynamics is crucial to the present phenomena as discussed later.

First, we treat the nuclear dynamics classically to clarify the quantum effects. For the qth harmonic generation, by using the steepest descent method, the induced dipole moment in a linear molecule whose axis is oriented at an angle of θ with respect to the x axis of the laboratory system is given by

$$\mathbf{x}^i_\theta(q) = \frac{2\pi i}{\sqrt{\det(S^i_{\rm e})''}}\left[\frac{2\pi}{i\tau_{\rm s}}\right]^{\frac{3}{2}} {\rm e}^{-iS^i_{\rm e}+iqt_{\rm s}}\mathbf{d}^*[\mathbf{p}'_{\rm s}, R^i_{\rm cl}(\tau_{\rm s}),\theta]$$

$$\times\mathbf{E}(t'_{\rm s})\cdot\mathbf{d}[\mathbf{p}_{\rm s}-\mathbf{A}(t'_{\rm s}),R_0,\theta]\,. \tag{4.9}$$

Here $\mathbf{p}_{\rm s} = \int_{t'_{\rm s}}^{t_{\rm s}} {\rm d}t''\mathbf{A}(t'')/\tau_{\rm s}$ with vector potential $\mathbf{A}(t'')$ and $\mathbf{d}(\mathbf{p},R,\theta) := \langle\mathbf{p}|\hat{\mathbf{x}}|\psi_0(\mathbf{R})\rangle$ is the bound-free dipole transition matrix element between the ground state $|\psi_0(\mathbf{R})\rangle$ and the continuum state $|\mathbf{p}\rangle$. Treating that the active electron in the step 2 feels the homogeneous but time-dependent potential, $S^i_{\rm e} = S^i_{\rm e-n} + [\tau_{\rm s} - 2(1-\cos\tau_{\rm s})/\tau_{\rm s} - C(\tau_{\rm s})\cos(2t_{\rm s}-\tau_{\rm s})]U_{\rm p}$ with the ponderomotive potential $U_{\rm p}$ and $C(\tau) := \sin(\tau) - 4\sin^2(\tau/2)/\tau$. $\det(S^i_e)''$ is the determinant of the 2×2 matrix of the second derivatives of S^i_e with respect to $t_{\rm s}$ and $t'_{\rm s}$. Here the parameter of electron acceleration [26] is set to be $\xi = 0$, following the numerical model by Lein et al. [33,34]; $\mathbf{p}'_{\rm s} = \sqrt{2q}(\mathbf{p}_{\rm s}-\mathbf{A}(t_{\rm s}))/|\mathbf{p}_{\rm s}-\mathbf{A}(t_{\rm s})|$.

For H_2 and D_2, the valence orbital $1\sigma_g$ is simply given by

$$\langle\mathbf{x}|\psi_{1\sigma_g}(\mathbf{R})\rangle = N'[\phi_{1s}(\mathbf{x}+\mathbf{R}/2)+\phi_{1s}(\mathbf{x}-\mathbf{R}/2)]\,,$$

where $\phi_{1s}(\mathbf{x})$ is the atomic $1s$ orbital in the configuration space, N' is the normalization factor, and $\pm\mathbf{R}/2$ are the positions of the nuclei. Then the dipole transition moments for H_2 and D_2 are given by

$$\mathbf{d}_{1\sigma_g}(\mathbf{p},R,\theta) = N'\big[2i\cos(\mathbf{p}\cdot\mathbf{R}/2)\mathbf{d}_{1s}(\mathbf{p}) - \sin(\mathbf{p}\cdot\mathbf{R}/2)\tilde{\phi}_{1s}(\mathbf{p})\mathbf{R}\big]\,, \tag{4.10}$$

where $\mathbf{d}_{1s}(\mathbf{p})$ is the atomic dipole moment from the $1s$ orbital and $\tilde{\phi}_{1s}(\mathbf{p})$ is the $1s$ wavefunction in the momentum space. The first term in (4.10) is the product of two parts and its physical meaning can be understood by the simple two-point emitter model with $\mathbf{p} = \mathbf{p}'_{\mathrm{s}}$. In fact, this term is zero when

$$\mathbf{p}'_{\mathrm{s}} \cdot \mathbf{R}/2 = (n + 1/2)\pi \Leftrightarrow R\cos\Theta = (n + 1/2)\lambda_{\mathrm{s}} \,,$$

where n is an integer and $\lambda_{\mathrm{s}} = 2\pi/p'_{\mathrm{s}}$ is the wavelength of the recombining electron with $p'_{\mathrm{s}} := |\mathbf{p}'_{\mathrm{s}}|$, and $\Theta := \arccos(\mathbf{p}'_{\mathrm{s}} \cdot \mathbf{R}/p'_{\mathrm{s}}R)$. The second term in (4.10), which is oriented along the molecular axis, corrects the error of the two-point emitter model [26].

Considering that $I_{\mathrm{p}}^{\mathrm{H}_2} \approx I_{\mathrm{p}}^{\mathrm{D}_2}$ and that the ionization rates at $R = R_0$ have little θ dependence [35], the dominant difference should be due to the electron and nuclear dynamics in the steps 2 and 3. The observed single molecule responses are obtained by superposing the radiation from all the molecules with random orientation coherently. For linearly polarized laser field, whose direction is defined as x axis, the observed dipole moment is given by

$$\begin{aligned} |\langle\langle x_\theta(q)\rangle\rangle| &\propto |\langle\langle d_x^*(p'_{\mathrm{s}}, R_{\mathrm{cl}}^i(\tau_{\mathrm{s}}), \theta)\rangle\rangle| \\ &= 2N'|\,\mathrm{sinc}(p'_{\mathrm{s}}R_{\mathrm{cl}}^i(\tau_{\mathrm{s}})/2) d_{x,1s}(p'_{\mathrm{s}})|. \end{aligned}$$

Here, $\langle\langle\cdots\rangle\rangle$ shows the average over all directions and $\langle\langle\sin(\mathbf{p}'_{\mathrm{s}} \cdot \mathbf{R}_{\mathrm{cl}}(\tau_{\mathrm{s}})/2)\rangle\rangle$ can be neglected for small R. Considering that the $|\,\mathrm{sinc}(p'_{\mathrm{s}}R_{\mathrm{cl}}^i(\tau_{\mathrm{s}})/2)|$ is a decreasing function of $R_{\mathrm{cl}}^i(\tau_{\mathrm{s}})$ in $p'_{\mathrm{s}}R_{\mathrm{cl}}(\tau_{\mathrm{s}})/2 \in [0, \pi]$, the intensity ratio $r_{\mathrm{CM}}(q) := |\langle\langle x_\theta^{\mathrm{D}_2}\rangle\rangle / \langle\langle x_\theta^{\mathrm{H}_2}\rangle\rangle|^2$ from classical treatment of nuclei satisfies

$$r_{\mathrm{CM}}(q) \approx \left| \frac{\mathrm{sinc}(p'_{\mathrm{s}}R_{\mathrm{cl}}^{\mathrm{D}_2}(\tau_{\mathrm{s}})/2)}{\mathrm{sinc}(p'_{\mathrm{s}}R_{\mathrm{cl}}^{\mathrm{H}_2}(\tau_{\mathrm{s}})/2)} \right|^2 > 1. \tag{4.11}$$

The ratio explains, however, only $\lesssim 30\%$ of the enhancement of HHG in D_2 ($r_{\mathrm{CM}}(q)$ in Fig. 4.8a): the remaining part is due to the quantum effects of nuclear dynamics.

In the full quantum mechanical frame, the valence orbital itself fluctuates due to the fluctuation of the nuclei, which forbids us to write R as arguments of states. Accordingly, the initial and final kets are given by $|\psi_{1\sigma_g}\rangle|w^i(t'_{\mathrm{s}})\rangle$ and $|\psi_{1\sigma_g}\rangle|w^i(t_{\mathrm{s}})\rangle$, and after inserting the identity operator of $\mathbf{1} = \int_0^\infty \mathrm{d}R|R\rangle\langle R|$ into (4.9), we obtain the final expression for the intensity ratio

$$r_{\mathrm{QM}}(q) = |s^{\mathrm{D}_2}(q)/s^{\mathrm{H}_2}(q)|^2, \tag{4.12}$$

where $s^i(q) := \int_0^\infty \mathrm{d}R w^i(R, \tau_{\mathrm{s}})^* w^i(R, 0)\,\mathrm{sinc}(p'_{\mathrm{s}}R/2)$ [27, 32] is the overlap integral. As can be seen in Fig. 4.8a, the experimental results are successfully reproduced with $r_{\mathrm{QM}}(q)$, which is reasonably determined by $R_{\mathrm{cl}}^i(\tau_{\mathrm{s}})$ and the quantum fluctuation $\delta R^i(\tau)$. However, $r_{\mathrm{QM}}(q)$ takes a rather complicated form and some assumptions such as GA [27] are needed to retrieve the nuclear motions.

On the other hand, as stressed in the introduction, the information on the dynamics of the nuclei is naturally encoded in the harmonic phases. In fact, the measured relative phase of harmonics is nothing but the difference of the action, and after expanding $I_{\rm p}(R) \approx I_{\rm p}(R_0) + (R - R_0)I'_{\rm p}(R_0)$, we obtain a simple expression for the relative phase,

$$\Delta\varphi(q) \approx I'_{\rm p}(R_0)\Delta R^{\rm H_2}(\tau_s)\tau_s/6, \tag{4.13}$$

where we used $\Delta R^{\rm D_2}(\tau) \approx \Delta R^{\rm H_2}(\tau)/2$. Finally, by solving this equation for $\Delta R^{\rm D_2}$, we obtain a simple relation between the position of nuclei and measured harmonic phase:

$$\Delta R^{\rm H_2}(\tau_{\rm s}(q)) \approx \frac{6\Delta\varphi(q)}{I'_{\rm p}(R_0)\tau_{\rm s}(q)}, \tag{4.14}$$

where $I'_{\rm p}(R_0) \approx 0.21$ a.u. and $\tau_{\rm s} = \tau_{\rm s}(q)$ are evaluated by a molecular orbital method and by following the procedure in [27], respectively. Notice that this relation (4.14) is derived from the only knowledge of the well-known ground state. Thus measuring the relative phases directly corresponds to observing the nuclear motions. Figure 4.8b shows the final results of the retrieved nuclear motion of H_2, $\Delta R^{\rm H_2}$ (squares vs. right axis) as a function of time τ (superior axis). The retrieved nuclear motion is consistent with the results of [27] and is successfully reproduced by our calculation (thick line) for $\Delta R^{\rm H_2}$ with the same method as [32].

4.5 Conclusion

In summary, we investigated HHG in mixed gases both experimentally and theoretically. As the new nonlinear media for HHG, using mixed gases can serve as a new route not only to control and characterize harmonics but also to observe attosecond dynamics in atoms and molecules.

Acknowledgements

This research was partially supported by the Ministry of Education, Science, Sports and Culture, Grant-in-Aid for Young Scientists (B), 20740235, 2008 and by The Special Post-doctoral Researchers' Program of RIKEN.

References

1. M. Hentschel, et al., Nature (London) **414**, 509 (2001)
2. M. Drescher, et al., Nature (London) **419**, 803 (2002); P. Tzallas, et al., Nature (London) **426**, 267 (2003); T. Sekikawa, et al., Nature (London) **432**, 605 (2004)

3. Y. Nabekawa, et al., Phys. Rev. Lett. **94**, 043001 (2005); **97**, 153904 (2006)
4. Y. Nabekawa, et al., Phys. Rev. Lett. **96**, 083901 (2006)
5. G. Sansone, et al., Science **314**, 443 (2006)
6. M. Uiberacker, et al., Nature (London) **446**, 627 (2007)
7. E.J. Takahashi, et al., IEEE J. Sel. Top. Quant. Electron. **10**, 1315 (2004); Appl. Phys. Lett. **84**, 4 (2004)
8. P.B. Corkum, Phys. Rev. Lett. **71**, 1994 (1993); J.L. Krause, et al., *ibid.* **68**, 3535 (1992); M. Lewenstein, et al., Phys. Rev. A **49**, 2117 (1994)
9. P.M. Paul, et al., Science **292**, 1689 (2001); S.A. Aseyev, et al., Phys. Rev. Lett. **91**, 223902 (2003); R. López-Martens, et al., Phys. Rev. Lett. **94**, 033001 (2005)
10. T. Sekikawa, et al., Phys. Rev. Lett. **88**, 193902 (2002)
11. T. Kanai, E.J. Takahashi, Y. Nabekawa, K. Midorikawa, Phys. Rev. Lett. **98**, 153904 (2007)
12. T. Kanai, E.J. Takahashi, Y. Nabekawa, K. Midorikawa, New J. Phys. **10**, 025036 (2008); Phys. Rev. A **77**, 041402(R) (2008)
13. R. Bartels, et al., Nature (London) **406**, 164 (2000)
14. P.R. Feynman, Rev. Mod. Phys. **20**, 367 (1948)
15. P.R. Feynman, P. Hibbs, *Quantum Mechanics and Path Intergrals* (McGraw-Hill, New York, 1965)
16. A. Rundquist, et al., Science **280**, 1412 (1998)
17. E.J. Takahashi, T. Kanai, K.L. Ishikawa, Y. Nabekawa, K. Midorikawa, Phys. Rev. Lett. **99**, 053904 (2007)
18. K. Ishikawa, Phys. Rev. Lett. **91**, 043002 (2003)
19. A.D. Bandrauk, et al., Phys. Rev. A **66**, 031401 (2002)
20. Y. Mairesse, et al., Science **302**, 1540 (2003)
21. C. Corsi, et al., Phys. Rev. Lett. **97**, 023901 (2006)
22. M. Lewenstein, et al., Phys. Rev. A **52**, 4747 (1995)
23. E. Constant, et al., Phys. Rev. Lett. **82**, 1668 (1999)
24. P. Salières, et al., Science **292**, 902 (2001)
55. J. Itatani, et al., Nature (London) **432**, 867 (2004)
26. T. Kanai, et al., Nature (London) **435**, 470 (2005); J.P. Marangos, Nature (London) **435**, 435 (2005); M. Lein, et al., Phys. Rev. Lett. **88**, 183903 (2002); T. Kanai, et al., Phys. Rev. Lett. **98**, 053002 (2007)
27. S. Baker, et al., Science **312**, 424 (2006)
28. H. Niikura, et al., Nature (London) **421**, 826 (2003)
29. C. Iaconis, I.A. Walmsley, Opt. Lett. **23**, 792 (1998); IEEE J. Quant. Electron. **35**, 501 (1999)
30. N.L. Wagner, et al., Proc. Nat. Acad. Sci. U S A **103**, 13279 (2006)
31. C.J. Latimer, et al., J. Chem. Phys. **102**, 722 (1995)
32. M. Lein, Phys. Rev. Lett. **94**, 053004 (2005)
33. M. Lein, et al., Phys. Rev. Lett. **88**, 183903 (2002)
34. M. Lein, et al., Phys. Rev. A **66**, 023805 (2002)
35. X.M. Tong, Z.X. Zhao, C.D. Lin, Phys. Rev. A **66**, 033402 (2002)

5

Propagation of Ultrashort Pulses in Condensed Media

Aditya K. Dharmadhikari and Deepak Mathur

Summary. This chapter summarizes some recent work on the propagation of intense, ultrashort laser pulses through condensed media. Propagation of ultrashort pulses results in formation of filaments and gives rise to supercontinuum generation. In the case of condensed matter like BaF2 crystals, six-photon absorption induced fluorescence is utilized to facilitate direct visualization of such femtosecond filaments. Some measure of control on the onset of filamentation can be achieved by changing the laser energy as well as laser polarization. Multiple filaments can be controlled in very simple manner, by using a wire mesh. Supercontinuum generation in transparent materials under various experimental conditions is summarized. Recent advances in application of such filamentation and white light generation in biological media as well as in material modification are described.

5.1 Introduction

Ready availability of femtosecond lasers has made it possible to carry out wide-ranging studies on the propagation of ultrashort pulses of light through transparent media. This area of research now constitutes an important subset of contemporary efforts on the part of physicists, materials scientists, and chemists to gain insights into how intense pulses of electromagnetic radiation propagate through matter. Such interest is driven both by fundamental considerations involving basic science [1–3] as well as by prospects of tantalizing applications in diverse areas ranging from the control of atmospheric lightning [4–6] and remote sensing [7], on the one hand, to atmospheric monitoring of pollutants by means of broadband, time-resolved spectroscopy [8, 9], on the other.

Propagation of pulses of intense light through condensed media opens a plethora of new possibilities over and above those that derive from free-space propagation. At the same time, there is a price to be paid in that concomitantly, the condensed medium also presents much more complexity as far as physical descriptions of the gamut of processes that determine the propagation dynamics are concerned. Nevertheless, there are several compelling reasons for attempting to circumvent the difficulties that are presented, especially from the perspective of applications. For instance, there has been considerable recent interest in utilizing intense, ultrashort laser pulses in studies that aim to achieve optically induced bulk modification of transparent materials. When strong light induces multiphoton processes within materials that possess large band gaps, the transfer of energy from the incident optical field to the material is readily facilitated, leading to practical opportunities in bulk modifications [10] like formation of voids and micro-cracks that result from laser-driven micro-explosions, as well as from optical breakdown that is also readily initiated [11]. Change of refractive index may also be induced by high-density effects [12]. Processes such as the formation of color centers and laser-induced oxidation–reduction of ions of transition metals, heavy metals, and rare-earth elements [13] occur. The bulk modifications that result from such processes might prove to be of utility in applications like the formation of waveguides [14] and photonic crystals [15], and in the development of three-dimensional optical memories [16,17]. Controlled laser machining within bulk material, without alteration of the surface properties of the irradiated material, is another area of considerable contemporary interest.

It is not difficult to appreciate that it is of utmost importance to develop experimental and theoretical methodologies that enable the major physical factors that govern the overall propagation dynamics to be understood [2, 3, 18]. The two most spectacular manifestations of propagation effects involving ultrafast pulses of intense laser light are filamentation and white light generation (also known as supercontinuum generation). The former effect manifests itself within the medium, while the latter is visualized downstream of the medium. We shall address facets of both these manifestations in the following and show that while much remains to be accomplished before a satisfactory level of insight is attained into the overall propagation dynamics in the strong-field, ultrashort regime, it is possible to utilize these propagation effects in potentially useful applications in diverse areas ranging from laser writing and micromachining to studies of interest in the life sciences and biomedical engineering.

The most prominent manifestation of propagation effects as intense short-duration pulses traverse optical media is the spectral broadening of the pulse, resulting in the production of a supercontinuum. During nonlinear propagation, such broadening is, initially, largely a consequence of self phase modulation. At very high values of incident laser power, it is known that plasma formation occurs. Generation of free electrons by multi-photon induced ionization processes makes significant new contributions to the spectral transformation, giving rise to broadening that is distinctly asymmetric. The high laser intensities that are the necessary starting point for supercontinuum generation are initiated by self-focusing collapse of the incident optical pulse within the medium. Such collapse has to be accompanied by a mechanism like multiphoton ionization (MPI that arrests an intensity catastrophe within the medium. MPI accounts for two arresting mechanisms: (1) a direct energy loss is provided for the collapsing field, and (2) plasma formation occurs within the medium; the plasma serves to defocus and absorb the increasing intensity. Plasma formation acts as a balance against the self-focusing effect: it limits the beam diameter at the self-focus region and, hence, limits the peak intensity that can be attained within the medium. In air, this "intensity clamping" [19] occurs along a filament at a peak intensity of about $5 \times 10^{13}\,\mathrm{W\,cm^{-2}}$ [20]. In general terms, MPI effects, along with plasma defocusing, clamp the maximum value of intensity, $I_{\max}$, that may be reached by a collapsing pulse. If ρ denotes the plasma density, it is clear that $\partial_t \rho$ will be proportional to $I^k_{\max}$, where k denotes the order of the multiphoton process; the value of $I_{\max}$ limits the maximum rate of plasma generation. Prevailing wisdom maintains that the major contribution to the blue side of the supercontinuum is the plasma-induced effects [21, 22], with the maximum blue-side frequency being directly proportional to $\partial_t \rho$.

Thus, it is the arrest of the maximum self-focusing intensity by MPI and plasma defocusing effects that governs the spectral extent of the supercontinuum that is generated. Recent numerical work [23] has offered indications that linear chromatic dispersion within the medium also plays an important role in determining the extent of spectral broadening that is achieved. The interplay between self-focusing is induced by the intensity-dependent refractive index of the propagating medium and plasma-induced defocusing that is brought about by multiphoton processes that determine the dynamics of femtosecond laser pulse propagation in optical media. From an experimental viewpoint, an important parameter in such dynamics is the minimum laser power that is required for self-focusing to occur within a given medium. The critical power for self-focusing, P_{cr}, is given in the continuous wave (cw) limit by [2]

$$P_{\mathrm{cr}} = 3.72\lambda_0^2/8\pi n_0 n_2, \tag{5.1}$$

where n_0, n_2 denote, respectively, the linear and nonlinear (intensity-dependent) refractive indices of the medium, and λ_0 is the incident laser wavelength. For infrared light ($\lambda_0 \sim 800\,\mathrm{nm}$) propagating through air, the value of P_{cr} is of the order of 3 GW [24]. For propagation through condensed media like silica, the corresponding value is much less ($P_{\mathrm{cr}} = 2.2\,\mathrm{MW}$).

The role of incident power in determining the complexity of the overall propagation dynamics is best illustrated with reference to recent numerical studies of (2+1)-dimensional propagation of femtosecond pulses through silica [24]. These studies were performed at incident power values that corresponded to $\sim 500 P_{\mathrm{cr}}$; the axially symmetric, extended, nonlinear Schrödinger equation was solved for the optical field, taking cognizance of the Kerr nonlinearity, group velocity dispersion, plasma formation, defocusing, self-steepening, and spatio-temporal focusing. Results of the simulations indicated that in the course of propagation through silica, the high power pulses appear to become spatially and temporally split into multiple cones as a result of the interplay of strong self-focusing and plasma defocusing. Whether this occurs or not depends very strongly on the incident laser power, P. When $P < P_{\mathrm{cr}}$, the low-power propagation is predominantly linear. When P is several times

P_{cr}, nonlinear effects start to become important and processes set in that give rise to filamentation and temporal splitting. In the very high power regime, when P is hundreds of times larger than P_{cr}, nonlinear effects dominate the dynamics and a number of complex effects have to be taken cognizance of, like multiple cone formation [24] and multiple filamentation [25].

Experimental confirmation of the formation of cone-like structures has been obtained in femtosecond time-resolved optical polarigraphy studies of induced instantaneous birefringence that occurs when high power (1–9 mJ), 130 fs long pulses of 806 nm wavelength propagate through material like quartz [10]. These experiments were conducted at power levels of ~1 GW, at values that are nearly three orders of magnitude larger than P_{cr} for quartz. These polarigraphy results showed that self-focusing can be an effective enough mechanism to induce absorption of a very large amount of optical energy at the local intensity maximum. As a result, a series of secondary areas of local intensity maxima are created, and these result in the formation of cone-like structure within the medium. In this high power regime, the spatio-temporal evolution of the laser beam within quartz is largely dominated by plasma defocusing and the nonlinear Kerr response. Processes like self-steepening and space–time focusing apparently exert only minimal influence on the laser beam profile. Note that in the low-power regime [26, 27], femtosecond continuum generation is enhanced by self-steepening of the pulse.

The duration of the incident laser pulse is another parameter that is under the control of experimentalists, and might prove to be of utility in helping disentangle some of the many effects that contribute to the overall propagation dynamics. In the long pulse regime, it has been known for more than three decades [28] that stimulated Raman scattering can deplete the pulse energy sufficiently to limit self-focusing. In contrast, in the regime of ultrashort laser pulses (pulses of less than 100 fs duration), self-focusing may also be limited by group velocity dispersion (by temporal stretching of the pulse) [29,30]. It is also clear that self-focusing results from the spatial distribution of nonlinear phase in a laser beam in much the same manner as self-phase modulation arises from the temporal distribution of nonlinear phase. In the ultrafast regime, SPM would be expected to be strong because the Stokes broadening is inversely proportional to the rise time of the laser pulse (conversely, the anti-Stokes broadening is inversely proportional to the pulse's decay time).

The implications of varying the experimentally accessible parameters on both on filamentation and supercontinuum generation within media are discussed in the following.

5.2 Propagation Effects: Filamentation

Femtosecond filamentation is a universal phenomenon observed in gases, liquids, and solids when the photon energy of incident laser pulse is smaller than the lowest energy required to achieve electronic excitation. There are many remarkable properties of filaments [1–3, 31], such as the fact that they are robust; they propagate over longer distances than the Rayleigh range; they are accompanied by processes like spectral broadening, the clamping of peak intensity inside the medium, conical emission, pulse self compression; and they give rise to improvement in spatial mode quality. Filaments are also known to break up into multiple filaments at higher values of incident power ($>>P_{cr}$). Here, we will restrict our discussion to propagation of ultrashort pulse through transparent solids and liquids.

Filamentation has been observed earlier in condensed media including transparent solids such as fused silica, water, and sapphire [32–35]. The long distance propagation of an ultrashort laser pulse in the form of a filament in transparent solid was obtained [32] in fused silica using a focused, 160 fs-long laser pulse. The filamentation exhibited nearly constant beam size that was maintained over many Rayleigh lengths. These filaments were observed along with supercontinuum generation, pulse splitting, and time compression [32, 36, 37]. We shall focus our attention on filamentation in transparent solids that are irradiated by very high values of incident power (>few hundred P_{cr}) and begin, naturally, by discussing the recent progress made in direct visualization of filamentation.

5.2.1 Visualization of Filamentation

Direct visualization of femtosecond filamentation is crucial to understanding the phenomenon. As the energy of a single infrared photon is much too small to effect an electronic transition, one has to take recourse to multiphoton absorption induced fluorescence to come up with a scheme to directly visualize filamentation in condensed media. One such scheme that has been successfully implemented involves the use of a crystal of barium fluoride, a material that is known to be very good scintillator [38].

Figure 5.1 depicts filamentation that has been imaged during the propagation of ultrashort pulses through a crystal of BaF_2. The spectrum associated with the filaments that are seen shows a predominantly blue emission [39] that is centered on 330 nm, one of the two luminescence bands (the other one

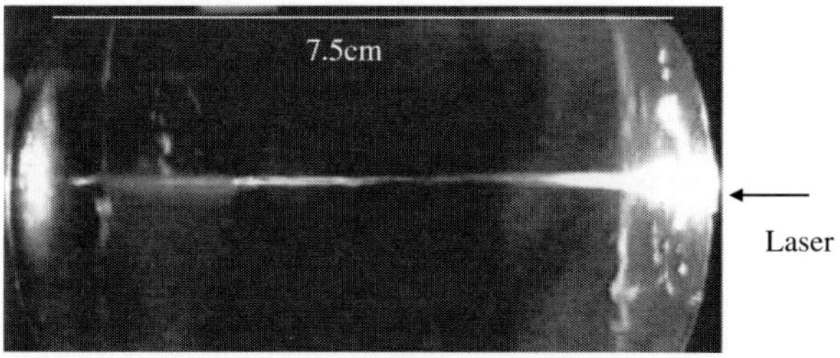

Fig. 5.1. Direct visualization of filamentation in a 7.5 cm long crystal of BaF_2. The incident laser beam enters from the right side

is centered at 200 nm) [38]. The band gap of BaF_2 is 9.2 eV, and as the photon energy of the incident radiation is 1.5 eV, the emission that is visualized in Fig. 5.1 is actually induced by a six-photon absorption process. Photons of 266 nm wavelength (the fourth harmonic of an ps Nd:YAG laser) have also been used to effect multiphoton absorption-induced emission in BaF_2 for visualization purposes [40] and the emission depicted a square dependence on incident laser intensity, indicating two-photon absorption process in this case.

5.2.2 Control Over the Onset of Filamentation

Self-focusing overcomes diffraction when the power of the laser beam exceeds P_{cr} and beam starts to collapse. The propagation length of the self-focusing beam, until the point at which collapse occurs, may be approximated by means of a semi-empirical formula [41]. In transparent media, collapse, or the onset of self-focusing, is amenable to control by varying the incident power of a collimated beam. In the case of external soft focusing (under the paraxial approximation), focusing conditions such as position of the sample and focal length of lens has to be taken into account [42]. It has also been shown [43] that by varying the laser power, the self-focusing distance in air can be altered.

Precise control over the onset of filamentation induced by femtosecond laser pulses has been achieved by changing both the laser energy and the beam divergence angle (the latter, by using a deformable mirror) [44]. Use of a telescope to increase the size of the input beam diameter and beam inhomogeneities [45] has enabled delay of the onset of self-focusing of the hot spots present in a terawatt laser pulse. The location of the nonlinear focus involving the use of a telescope has been derived by [46] and has found support from experiments. We now discuss two experimental situations to demonstrate control of the onset of filamentation. One is by changing the incident laser power and the second is by rotating the plane of polarization of incident light. Examples of condensed media in which such control has been demonstrate include crystals like BaF_2 (15 mm length) and sapphire (3 mm) [47]. In the case of BaF_2, we have already noted that filamentation can be readily monitored by imaging the fluorescence signal obtained upon six-photon absorption of incident laser light. Such visualization can be used to make direct measurements on control of the onset of filamentation within the crystal. By way of illustration, consider the case of soft external focusing, with the crystal-to-lens distance fixed such that the focus of the lens lies beyond the crystal position.

Figure 5.2 shows that the onset of collapse in barium fluoride as the incident power is varied from 110 P_{cr} to 1,200 P_{cr}. The onset of filamentation, z_{of}, the filament "start" position (the arrow in each panel marks the onset position), varies from 4.6 mm inside the crystal at 110 P_{cr} to a position that is located on the incident face of the crystal at 1,200 P_{cr}. Note that the images in Fig. 5.2 indicate that multiple filamentation occurs within the crystal at such high power levels.

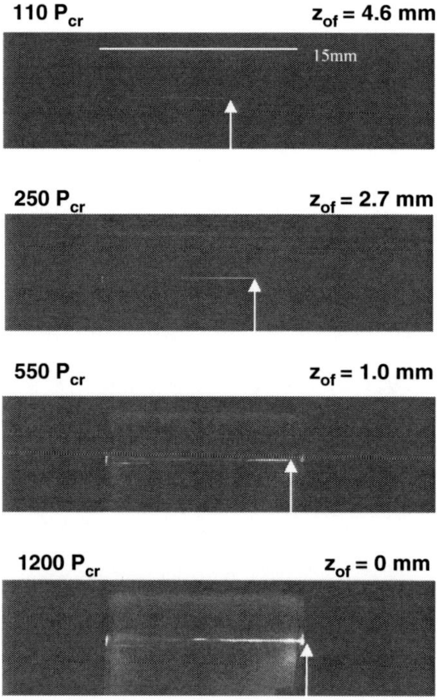

Fig. 5.2. Start of filamentation in BaF_2 at different values of incident laser power expressed in multiples of P_{cr}, the critical power for self-focusing. Vertical *arrows* indicate the "filament start" position in each case

Another illustration involves keeping a similar focusing condition and incident laser power (100 P_{cr}), and altering the direction of polarization of the incident laser light by means of a quarter-wave plate. It has been shown that z_{of} is a function of incident polarization angle [47]. We note that to obtain a similar change in z_{of} without use of polarization control, as much as a fourfold increase in incident laser power would be required. From the perspective of avoiding optically induced damage to irradiated materials, it is obvious that polarization control scores heavily over control that is exercised using incident power.

Material like sapphire, which is a uniaxial crystal, does not show nonlinear fluorescence, and direct visualization of filaments is, consequently, not an option. However, the effect of rotation of plane polarization can also be probed in such materials by monitoring the spatial size of the white light disc, and of the spectrum of the supercontinuum that is produced. Figure 5.3 depicts some typical results obtained with 3 mm long sapphire crystals. The extent of the supercontinuum spectrum is seen to vary as a function of polarization angle. The supercontinuum spectrum has two components [39]: symmetric broadening about the incident wavelength that is essentially ascribable to

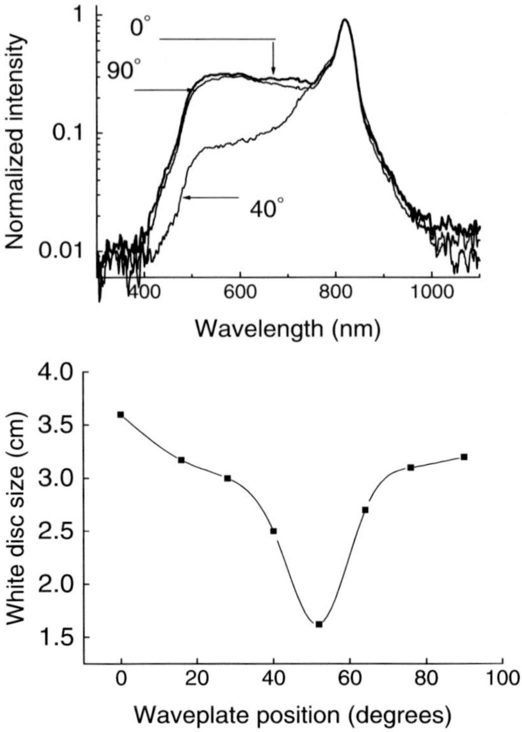

Fig. 5.3. *Top panel*: Polarization-dependent supercontinuum spectra obtained in sapphire. The corresponding variation in the diameter of the white light central spot as a function of incident polarization angle is shown in the *lower panel*

self-phase modulation effects and an asymmetric component that arises from self-steepening and MPI-induced free electron generation. The asymmetric contribution is on the blue side of the spectrum and exhibits strong polarization dependence [47]. Figure 5.3 also shows that the size of the white light disc has a distinct minimum at a certain value of polarization angle. These results may be rationalized in the following terms. In uniaxial crystals like sapphire, the relation for effective $\chi^{(3)}$ is not simple as there is direction dependence for extraordinary (e) rays but not for ordinary (o) rays. The value of P_{cr} will, consequently, vary accordingly. Incident ultrashort pulses undergo splitting in the sapphire crystal, with the e and o pulses propagating with different group velocities. Consequently, the splitting ratio depends on the incident polarization. The group velocity mismatch length is of the order of 3 mm [47]. The area under the curve of the white light spectra show a 30% reduction at an angle of 45° compared to that at 0° [47]. There is major polarization-dependent spectral change, specifically a reduction in the spectral width as the incident laser polarization nears 45°. We note the similarity in this functional dependence with that demonstrated for BaF_2.

The above discussion pertains to crystals possessing intrinsic birefringence (like sapphire) and induced birefringence (like BaF_2 see [47]). With the $\lambda/2$ plate at 0° or 90°, the white light spectra are more or less identical. At 45° position, the amplitudes of the e and o pulses are equal. Further, because of GVM-induced pulse splitting, each pulse undergoes self-focusing at different times but with half the incident power. As a result, one obtains a manifestation of GVM induced splitting, in that spectra measured at 0°, 20°, and 40° are very different from each other. The reduction in the blue component of the white light spectrum sensitively maps the decrease of effective laser intensity within the medium. This is closely related to changes associated with self-focusing conditions due to GVM-induced pulse splitting.

5.2.3 Focusing–Refocusing Events

We now consider the situation wherein a single filament propagates through some condensed medium and we monitor the different events that occur, particularly on sequences of focusing–defocusing–refocusing. A single filament is observed when relatively low incident laser powers are used, close to P_{cr}. Two or more filaments begin to appear as the incident power increases. We again make use of the 6PA fluorescence discussed in Sect. 2.1 to directly observe focusing–refocusing events within the crystal. Such measurements facilitate quantitative estimation of plasma parameters within a crystal, such as the electron density, local laser intensity, and the 6PA cross section.

We illustrate such capability by discussing results of experiments conducted using a Ti-sapphire laser (800 nm wavelength, 40 fs and 1 kHz repetition rate) to generate a single filament in a 15 mm long BaF_2 crystal. The incident laser beam was loosely focused onto the crystal such that the physical focus was located outside the crystal. Filament size was measured using a CCD camera attached to a long working distance objective lens [48]. Figure 5.4 shows the transverse view of a single filament observed at 5 μJ energy, with the diameter being 10 μm. With increase of incident energy to 8 μJ, a second filament was seen to appear, while at still higher energy (100 μJ) multiple filaments were observed. By imaging the white light continuum and conical emission corresponding to the single filament, an overall picture of propagation effects vis-a-vis filamentation and supercontinuum generation are obtained.

Figure 5.4 also shows a typical image of the conical emission and white light (central disc) emerging from the crystal in the laser propagation direction. It should be noted that the order of appearance of a given single color during conical emission is opposite to that which would be expected due to diffraction. A possible physical mechanism for such behavior has been discussed in an earlier volume in this series [1]. Many other mechanisms have been also proposed to describe conical emission, such as a Cerenkov type process [49], four-wave mixing [50], and X-waves [51], but it is probably fair to state that a clear understanding of the underlying physics is yet to emerge.

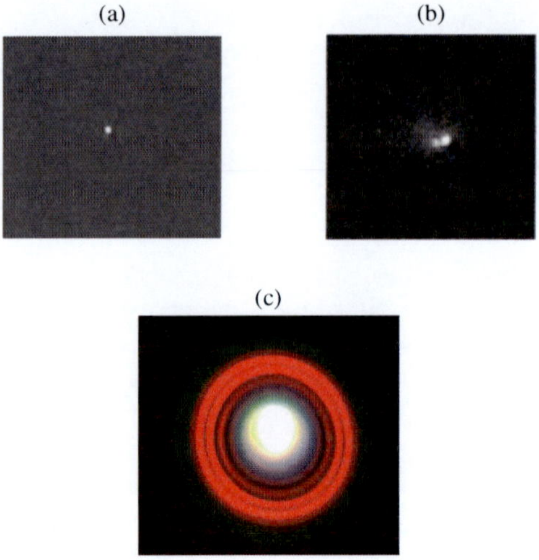

Fig. 5.4. Transverse view of (**a**) a single filament obtained upon irradiation of a 15 mm crystal of BaF_2 by incident energy of 5 μJ, (**b**) two filaments obtained at 8 μJ, and (**c**) the corresponding image of the conical emission and white light obtained in the far field

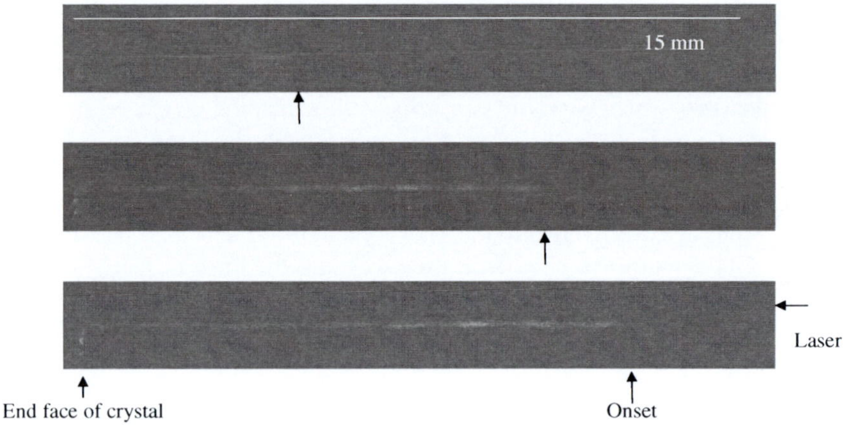

Fig. 5.5. Longitudinal images of filaments obtained at incident energies of (**top panel**) 10 μJ, 12 μJ (**middle panel**), and 15 μJ (**lower panel**)

Typical images of filamentation parallel to the laser propagation direction are shown in Fig. 5.5 for various values of incident power.

It is seen that at ∼10 μJ incident energy, there are two focusing events, with the distance between the two self-focusing events being ∼1.8 mm. As the energy is increased from ∼12 to ∼15 μJ, there is an increase in the number of focusing–refocusing events, with the corresponding spacing between the peaks decreasing to 1 mm. At still higher energy, merging occurs of the focusing–refocusing events inside the filament. The third and fourth refocusing events have higher fluorescence intensity compared to others events, indicating an increase in the intensity inside the filament at these locations. The nonlinear fluorescence signal depends on incident intensity I and scales as I^k, where k is the number of photons that are absorbed. Thus, this type of measurement is able to map the intensity inside the filament. The propagation length of the filament that is depicted is ~9 mm. The number of focusing–refocusing cycles depends on the peak power of the input pulse (Fig. 5.5).

Theoretical investigations of the self-guiding effect in condensed media have been carried [52] and it has been observed that the collapse of the beam in fused silica and water after self-focusing is arrested by multiphoton absorption at lower wavelengths as opposed to plasma defocusing. At higher wavelengths, plasma defocusing is the dominating mechanism to arrest self focusing. As the results depicted in Fig. 5.5 were from experiments carried out at 800 nm, it is probably correct to take plasma defocusing to be the dominant mechanism that arrests self-focusing. By measuring the filament radius ($L_{\min}$) inside the BaF_2, it becomes possible to estimate peak intensities ($I_{\max}$) and electron densities ($\rho_{\max}$) by taking into account diffraction, the Kerr effect, and ionization responses only. It should be noted that other effects, like group velocity dispersion, diffraction, self-phase modulation, and pulse self-steepening, may also contribute to the self-guiding process.

Following the treatment given earlier [2, 3, 53], one can estimate the peak intensity, electron density, and the six-photon absorption cross section from experimental observation of filament radius. With the size of the single filament (Fig. 5.4) taken as 8.3 μm (end-to-end diameter), the use of the above equations enable estimates to be obtained for the peak intensity ($I_{\max}$) and electron density within the crystal as $3.26 \times 10^{13}\,\mathrm{W\,cm^{-2}}$ and $2.81 \times 10^{19}\,\mathrm{cm^{-3}}$, respectively [48]. Using these values, it also becomes possible to estimate the 6PA cross-section in BaF_2 to be $0.33 \times 10^{-70}\,\mathrm{cm^{12}\,W^{-6}\,s^{-1}}$.

5.2.4 Other Control Issues: Multi-Filamentation

It is known that the spatial profile of ultrashort laser pulses does not have smooth envelope but usually has small structures that grow into separate filaments during propagation through a medium. This process has been described [54] as interference between a strong beam and weak beams due to the intensity-dependent refractive index. Such interference gives rise to index variations behaving like a grating. The strong beam redistributes energy into the weak beams due to the grating action. The formation of a filament is random in nature and, hence, is not amenable to control.

There are many potential applications of filaments (some of which have been discussed above) that require exercising control over the multiple filamentation process. We discuss some of the methods that might possibly be of utility in this respect. To control filamentation in air, [43] introduced amplitude masks at the input of the incident beam. Schroeder et al. [55] performed similar investigations for controlling filaments in water using a wire grid that was placed at the input of beam. Nonlinear fluorescence techniques were employed to visualize multiple filaments from a dye dissolved in methanol. The filamentation dynamics was observed via two-photon absorption-induced fluorescence. Several processes were identified, such as filament extinction, refocusing after a bright fluorescence, breakup of filament, and fusion of filaments. Other methods for filament control have utilized the effects of beam astigmatism and ellipticity [56,57]. It has been demonstrated [58] that diffraction from a slit can be utilized to control filament formation. Independent control of filament starting position and length of filament has also been achieved [59] by manipulating the spectral phase of the incident laser pulse as it propagates in water. Switching of white light patterns over femtosecond timescales has been attained using two laser pulses and controlling their local overlap in space and time [60].

Multiple filamentation at very high incident power compared to the critical power for self-focusing in CS_2 was observed by Centurion et al. [61] who also noted that beams can be made to break up into a constellation on which filaments can be made to form. It has also been shown [62] that in CS_2 the beam break-up into multiple filaments may occur even before the ionization intensity threshold is reached, while ionized electron plasma is detected in the media like water and air. We discuss in the following a relatively simple technique proposed by Schroeder for exercising control [63] by means of an ordinary wire mesh that produces an intensity distribution in the x–y plane and along the z-direction. Such intensity distributions due to interferences give rise to local intensity minima and maxima, with the latter serving as the seed for nonlinear processes to set in. Translation of the mesh is a simple way of exercising control. By way of illustration, consider mesh-based propagation in various condensed media [63] with 40 fs duration of 810 nm light of maximum energy 600 μJ and beam diameter of 3 mm. Consider specifically regularization of filamentation using the mesh (Fig. 5.6). The beam exiting the mesh propagates through a 5 cm cell containing coumarin dye in ethanol solution. Multiple filamentation is visualized by the fluorescence emitted by

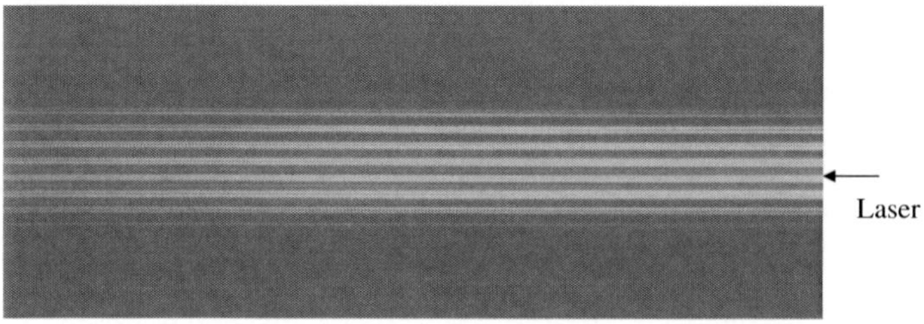

Fig. 5.6. Multiple filamentation in a 5 cm long glass cell containing coumarin dye in ethanol. Control over filamentation was obtained by means of a wire mesh (see text)

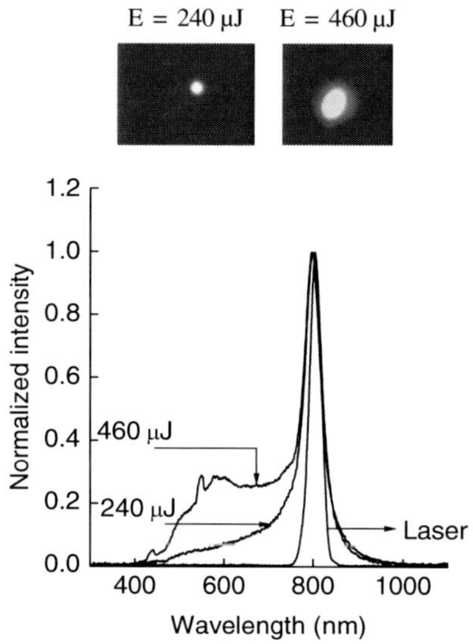

Fig. 5.7. Supercontinuum spectra obtained in water at different energy levels using a wire mesh. *Top panels* show images of white light in the far field

the dye molecule due to two-photon absorption of 810 nm radiation (Fig. 5.6). Filamentation is seen to occur throughout the 5 cm length of the cell and possesses a regular pattern.

Apart from visualization of multiple filamentation, broadband emission from filamentation in water is also readily observed [63]. Such broadband emission consists of two parts: a central portion, which is the on-axis component and is known as the supercontinuum or white light emission, and second component that is off-axis and has ring-like conical emission. The supercontinuum is generated using a wire mesh without using any external lens in material like water and barium fluoride with the incident energy in the range of 240–460 μJ. Figure 5.7 shows the spectra of supercontinuum generated for two different values of incident laser power.

Note the increase in the supercontinuum intensity towards the blue side of the spectra as the incident power increases. We shall discuss this in more detail in the following section. Images are also shown of white light generated at two energy values. Consider now the off-axis conical emission. Keeping in mind that the wire mesh acts like a local focusing element enables one to quantify the efficiency of conical (off-axis) emission. In contrast to a focusing geometry situation, by using a wire mesh the incident beam is kept essentially parallel at low intensities. We observe that above a threshold value of incident laser power, the beam exhibits extra divergence (conical emission), thereby reducing the on-axis intensity. The method described [63] provides a simple technique to measure efficiencies for the off-axis emission. Below a certain value of threshold energy, it is observed that conical emission is absent. At higher incident energies conical emission is observed to "switch on." Further, conical emission is that part of the incident beam energy that gets deflected (emission angle >2.5 mrad) and, thus, fails to fall on the detector area in the far field region. Measurements of the conical emission efficiency for different condensed media ranged from 16 to 87% [63] and reflect conservation of energy for nonlinear propagation of light. In case of water, benzene, and CS_2, the conical emission efficiency was 35%, 61%, and 87%, respectively. In glass and BaF_2, the efficiency was 42% and 47%, respectively. Materials with high nonlinear index coefficient yield higher conical emission efficiency [63]. Measurements carried out in water have established a close connection between CE and supercontinuum. Physical mechanisms like pulse-splitting, SCG, and CE result from a single phase matching process consisting of two split pumps and two X-waves [2, 64].

5.3 Propagation Effects: Supercontinuum Generation

The femtosecond supercontinuum is usually a visually spectacular effect and has been of interest for the past two and a half decades [26, 65]. The continuum appears as a white disk surrounded by a distinct, concentric, rainbow-like conical emission (Fig. 5.4); the low-divergence central part of the beam has come to be called the white-light continuum or supercontinuum. There exist various mechanisms that might possibly explain the supercontinuum generation (SG): self-phase modulation [65]; ionization-enhanced SPM [65, 66], with additional contributions from the interplay of diffraction and instantaneous electronic Kerr nonlinearity [67]; stimulated Raman scattering; self-steepening; and four-wave parametric processes [27, 68–71]. In spite of this plethora of mechanisms, a definitive understanding of the rich physics that is involved in SG is still lacking [72]. Based on earlier results on the picosecond continuum [73], the threshold power for SG was found to coincide with the calculated critical power for self-focusing. Thus, self-focusing plays an important role in the white light continuum generation [66, 74]. In this section we discuss results of SG in transparent materials under various conditions such as the physical focusing conditions, the pulse duration of the incident radiation, the polarization of the incident laser pulse, coherence and incident laser power, as well as the choice of material through which propagation occurs. We further discuss the efficiencies and the extinction ratio of the SG.

5.3.1 Material Property

The choice of material for SG is very important and was first reported a decade ago by Brodeur and Chin [21, 75] using a femtosecond laser in condensed media. They observed that spectral broadening of the white light depends on the band-gap of the irradiated material [75]. Furthermore, they found the existence of a band gap threshold, ~4.7 eV, below which a medium does not generate a supercontinuum. Above this threshold band gap value, the width of the continuum spectra was found to increase with band gap. The primary mechanism proposed for such behavior [21, 73, 75] was enhancement of free electrons that are generated by multi-photon ionization-induced SPM. MPI leads to plasma formation within the medium. The plasma then absorbs, defocuses, and spectrally blue-shifts the incident laser field. A dynamic competition between the Kerr effect and plasma defocusing leads to a clamping of the maximum intensity inside the medium that limits the spectral extent of SG [21]. This is the process known in the literature as intensity clamping. In addition to the band gap of the medium, chromatic dispersion [23] was also shown to contribute towards limiting the spectral extent of SG in transparent media. Further self-steepening was shown to be another process that significantly affected the spectral extent of SG [70]. The effect of both self-steepening and group velocity dispersion on supercontinuum generation has recently been reported in water [76]. In the following, we illustrate some of these features in relation to SG in materials with band gap much in excess of 4.7 eV, such as BaF_2, sapphire, water, and BK-7 glass.

5.3.2 Focusing Conditions

Most early work on supercontinuum measurements used incident power levels that were close to P_{cr}, wherein rings and conical emission constituted a significant portion of the total output signal. At very high power levels, it is the central white light portion that overwhelmingly dominates the total signal output [40]. By using loose focusing geometry and high incident power, SG with efficiency levels well in excess of ∼40% [40] have been demonstrated in barium fluoride. A representative spectrum is shown in Fig. 5.8.

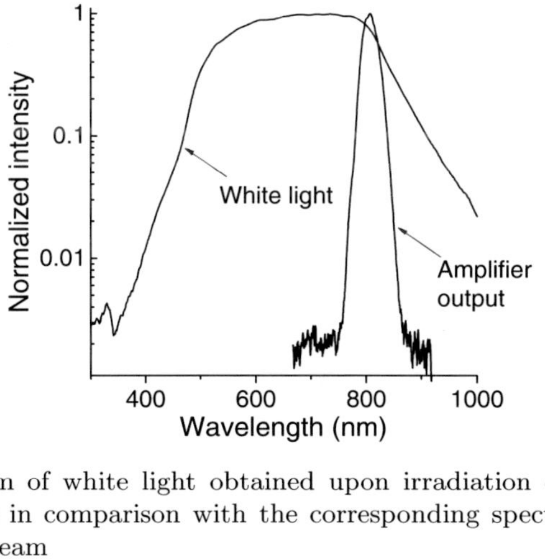

Fig. 5.8. Spectrum of white light obtained upon irradiation of BaF_2. Note the extent and flatness in comparison with the corresponding spectrum measured for the incident laser beam

To optimize the conversion efficiency, physical focusing conditions need to be optimized. With reference to the SG spectrum shown in Fig. 5.8, for fixed incident laser power and fixed focal length of the lens, the location of the focus point within the 7.5 cm long barium fluoride crystal was found to determine the extent of the white light spectrum. With the focus located 1 cm inside the crystal, the flattest possible spectrum of white light was obtained (as shown in Fig. 5.8), and the highest conversion efficiency was also achieved.

Altering the focus position such that it is located near the center of crystal (Fig. 5.9a) leads to a significant decrease in intensity in the region 400–500 nm. When the laser beam is focused well inside the medium, the intensity in the spectral region from 400–600 nm is again significantly reduced. These observations can be understood by considering dispersion that broadens the pulse duration of the laser, thereby reducing its intensity. The broadest white light spectrum is obtained with the focus very close to the front face of the crys-

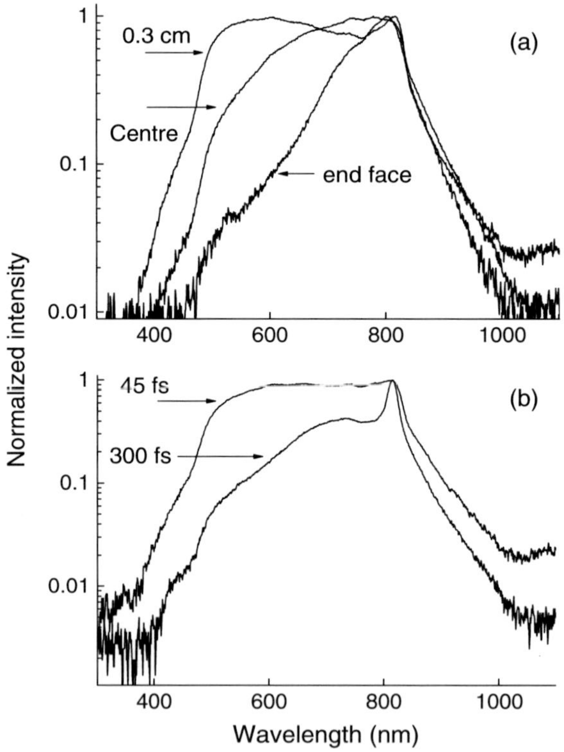

Fig. 5.9. (**a**) Spectra of white light obtained upon irradiation of BaF_2 as a function of the position of physical focusing within the crystal. 0.3 cm indicates that the focus lies fairly close to the entrance face of the crystal. Note how the extent and flatness of the spectrum depends on focus position. (**b**) Corresponding variation in spectral extent and shape obtained with incident pulses of 45 and 300 fs duration

tal, confirming that the dispersion, indeed, plays an important role. However, under such high intensity conditions, damage to the crystal face is a likely outcome [77].

5.3.3 Pulse Duration Dependence

We now consider the dependence of SG on the incident pulse duration, particularly at higher incident powers, in terms of spectra measured using two different Ti:sapphire lasers producing 45 and 300 fs pulses at the same laser power [77]. The resulting spectra (Fig. 5.9b) show a significant difference, particularly in the blue region of the spectrum. With 45 fs pulses, the spectrum is broad and conversion appears to be higher than in the case of 300 fs pulses. These observations may be rationalized using the fact that SPM is known to be inversely proportional to the pulse duration [26]. Furthermore, free electron generation contributes to SPM, particularly leading to the enhancement of the blue spectral region [21]. Thus, using pulses of two different durations seems to indicate that the free electron contribution to SPM is larger in the case of 45 fs pulses than for longer (300 fs) pulses. Furthermore, upon increase of the laser energy in the case of 300 fs pulses, severe damage to the medium occurs for energies in excess of 5 mJ, whereas no damage was found with 45 fs pulses at incident energies as large as 1 mJ.

5.3.4 Polarization Dependence

We now discuss issues regarding the dependence of SG on incident laser polarization and the polarization properties of the supercontinuum itself. Early experiments carried out on water with 100 fs pulses showed that SG is reduced when the incident light is circularly polarized [78]. Theoretically, it has long been known that the effective ionization energy for circularly polarized beams is higher than for linearly polarized light [79] and that multiphoton ionization is less efficient for circularly polarized light [80]. Numerical simulations have probed the polarization dynamics of femtosecond propagation in air [81]. Here also it has been observed that SG is more pronounced when linearly polarized light is used compared to the situation when an equally intense, circularly polarized light is used. We have already discussed that SG depends strongly on the incident laser power and it is noteworthy in this context that the critical power for self-focusing for circularly polarized light is 1.5 times that for linear polarization [81]. In a theoretical study involving multiple filamentation in a Kerr medium, a higher threshold for self-focusing was found for circularly polarized laser beam [82]. Consider experiments conducted in BaF_2 using linearly and circularly polarized laser beams of 45 fs duration of the same energy (450 μJ) to illustrate some important facets of the dynamics [77]. The intensity of white light is higher for linearly polarized light compared to that for circular polarized light and the blue side of the spectrum is broad and intense for the linearly polarized beam. We have already shown that generation of the blue part of the white light spectrum requires a higher intensity than that required for the central-wavelength region. Presumably, this is why the variation of the white light intensity with laser polarization becomes most significant in the blue region of the spectrum. Such difference in SG may also be rationalized by considering both SPM processes and MPI-enhanced SPM. The free electrons produced by MPI enhance SPM and contribute towards the generation of the blue side of the supercontinuum. It is known that the effect of light polarization on the multiphoton ionization of atoms is related to the field correlations of multiphoton processes [83]. Theses effects arise due to the fact that the vectorial nature of the incident radiation field affects the transition amplitudes of multiphoton processes. Cross products of matrix elements involving the orthogonal components of the radiation field occur in the expression of the transition amplitude for circularly polarized light, and these give rise to the dependence of ionization rate on the polarization state of the incident light field.

We now discuss how SPM processes contribute towards the polarization dependence of white light generation [77]. We note that our illustrative sample, BaF_2, is a cubic crystal with inverse symmetry; it is optically isotropic and does not show linear birefringence at 800 nm. But when the crystal is exposed to the intense field of ultrashort laser pulses, SPM and cross phase modulation (XPM) give rise to field-induced change in the nonlinear refractive index that is dependent on the incident intensity [84, 85]. Through the XPM process, orthogonal field components couple in nonlinear fashion. The change in nonlinear refractive index is unequal for two orthogonal crystal axes as it depends on both the polarization and the intensity of the laser light.

We now turn our attention to the polarization properties of the generated supercontinuum. The earlier experiments on white light generation in isotropic media by Alfano and Shapiro [86] indicated that the polarization of white light is same as that of the incident laser. To get further insight, polarization is quantified by measuring extinction ratio (ER) of supercontinuum generation ($I_{\|}/I_{\perp}$), which denotes the ratio of intensity transmitted when an analyzer is parallel to the polarization of the laser to when the analyzer is in crossed or perpendicular position. The ER of SG in glass has been found to decrease as a function incident power. At low incident power, the ER of the supercontinuum is the same as that of the incident laser (nearly 400:1). As the incident power is increased, the ER reduces and at incident power close to 1,000 P_{cr} it reduces by nearly two orders of magnitude [87]. The reduction in ER is not yet completely understood, but it is likely to be a consequence of the fact that multiple filamentation that is known to occur at this incident power level may lead to a complicated refractive index change, giving rise to local anisotropy that degrades the uniformity of polarization. The depolarized behavior of white light has been recently confirmed in the experiments involving fragmentation of methane molecule by intense white light [88].

5.3.5 Coherence

Supercontinuum generation is the result of many competing processes and, consequently, coherence is an extremely useful property to consider in understanding how the phase relation is maintained during its generation by mode-locked lasers. The spatial coherence can be determined in a simple way by Young's double slit geometry. For fringes to be formed, at least two sources of white light are required. As already noted, it is established that above a certain threshold power level, the input beam breaks into small filaments inside a medium. Each filament then generates white light continua acting as a source for interference. The regime of high incident peak powers has been experimentally investigated [89] and it is known that the beam breaks up into randomly distributed small-scale filaments, each of which acts as a source of white light. Early experiments showed that each spectral component of the white light acts as a coherent source [89]. It has recently been shown that white light generated from multiple filaments also exhibits a high degree of spatial coherence [90]. White light supercontinuum interference of self-focused filaments in water has been probed using cylindrical lenses [91] and a stable one-dimensional array of filaments has been generated that gives rise to interference effects between adjacent filaments. The stable interference pattern created by a filament pair is similar to that obtained in a conventional Young's double slit set-up, in agreement with predictions that the filaments are spatially coherent sources. This implies a constant phase relationship between the supercontinua generated by each of the many filaments. Subsequent work has confirmed that this interference [40] can be observed using both spherical and cylindrical lens focusing.

Figure 5.10 shows images of filaments and the corresponding interference patterns obtained upon irradiation of BK-7 glass by low incident powers using both cylindrical and spherical lens. The advantage of using a cylindrical lens is that one can generate a horizontal array of stable white light sources. Similar results have been obtained [92] in a linear array of phase coherent supercontinuum sources.

5.3.6 Incident Power Dependence

Supercontinuum generation obtained at different values of incident power might help disentangle the effects of plasma formation from those that are SPM-induced. As such measurements also involve material being exposed to high incident power levels, it requires close monitoring for any obvious damage. By way of illustration, consider 42 fs pulses propagating through 7.5 cm of BaF_2 at incident power levels ranging from 90 P_{cr} to 3,600 P_{cr} [39]. Loose focusing conditions of the incident laser beam are maintained. Figure 5.11 shows typical broadband spectra. A nearly flat spectrum is generated when incident power levels are about 2,000 P_{cr} and above, with high conversion efficiency values (>60%) being obtained.

At lower power values, there is relatively less spectral broadening, and it appears to be mostly symmetrical around the incident wavelength of 801 nm,

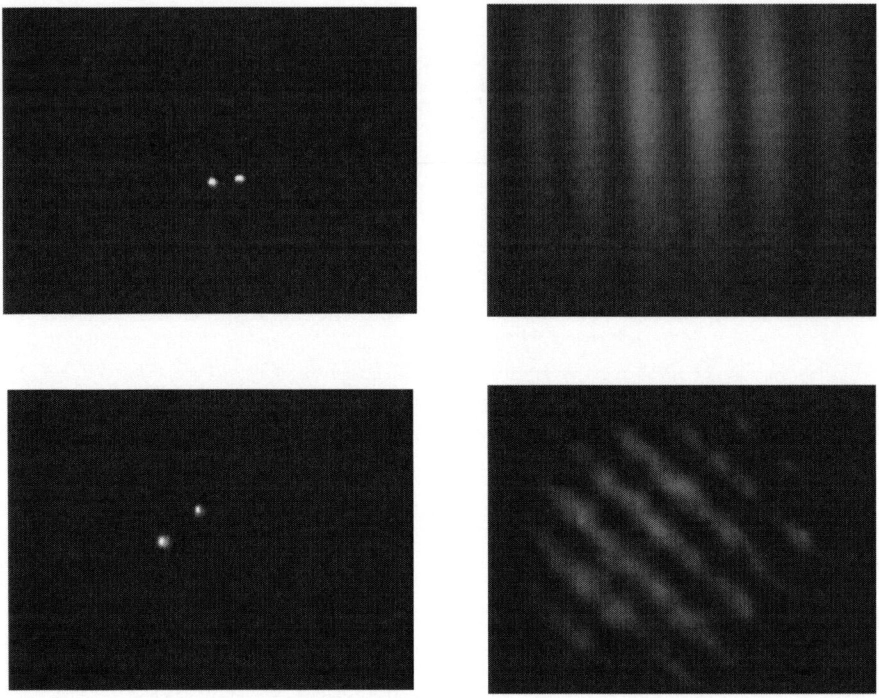

Fig. 5.10. The top two panels show the transverse image of two filaments (*top left panel*) and the corresponding interference pattern (*top right panel*) observed in the far field obtained using a cylindrical lens. The *lower panels* show the corresponding images when a spherical lens is used

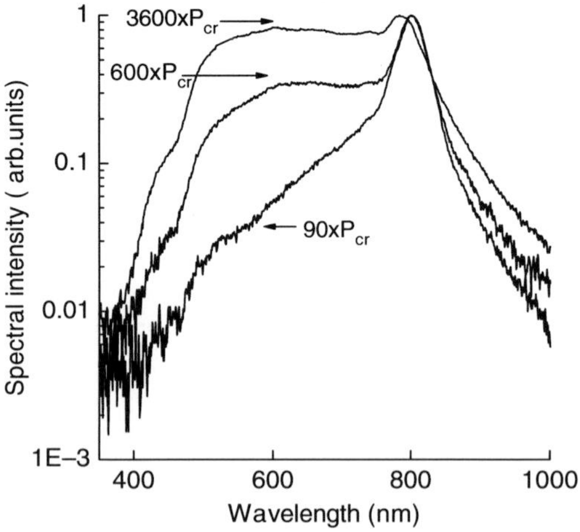

Fig. 5.11. White light spectra obtained upon irradiation of BaF_2 at different incident power levels

in contrast to the situation at higher powers when a pronounced asymmetry begins to set in such that at 3,600 P_{cr}, the contribution from 801 nm region is only marginal [39]. Another interesting observation at the higher powers is blue-shifting the peak of the white light spectrum. At low power levels, the spectral broadening of the ultrashort pulse is due essentially to SPM. In the leading part of laser pulse, generation of lower frequencies (Stokes broadening) takes place, whereas in the trailing part of the pulse there is just the opposite situation, giving rise to higher frequencies (anti-Stokes broadening). The frequency deviation is symmetric relative to the peak of the frequency distribution of the incident laser pulse. With increase of the incident laser power, apart from the Kerr effect, free-electrons that are generated due to MPI also begin to contribute to spectral broadening. The above observations are consistent with simulations [93] that take into consideration contributions of the refractive index from the Kerr nonlinearity and the plasma. On the basis of recent results [10], it appears that in the power regime that lies a thousand-fold, or more, above the value of P_{cr} for a given optical medium, contributions from processes like self-steepening are essentially insignificant. In the high power regime, the time variations of the refractive index due to contribution of free electrons give rise to asymmetric spectral broadening that is blue shifted, while the space variation of the refractive index gives rise to defocusing. The free electrons generated due to MPI produces plasma that induces a spectral shift. The plasma-induced spectral shift, $\delta\lambda$, with respect to the incident laser wavelength analyzed using a homogeneous model that assumes plasma being described by a single-valued refractive index n [94] that can be related to the plasma length, the incident laser wavelength, and the free-electron density. Note that no shift is observed in the incident laser wavelength (801 nm) at incident powers as high as 6 GW. Beyond this, a measurable blue shift of 17 nm is observed at 12 GW, which corresponds to about 3,000 P_{cr}. Further increase in the laser power, to 3,600 P_{cr} (14 GW), results in a larger blue shift, of about 20 nm [39]. One can deduce time varying free-electron density if the plasma length can be established. Though it is difficult to quantify the plasma length, however, an estimate can be obtained. The length of the filaments that are imaged ~3 cm and at the high power levels, it is most likely that multiple filamentation occurs. Nevertheless, an order of magnitude estimate of the time varying free-electron density for plasma lengths in the range from 1 to 3 cm, are in the range 2–7 $\times$ 10^{25} $\text{cm}^{-3}\,\text{s}^{-1}$ [39].

5.4 Applications of White Light Generation and Filamentation

There are numerous applications of white light generation and filamentation that span across disciplines, such as time-resolved broadband spectroscopy, generation of few cycle pulses [37], LIDAR [31], material modification [2], and bio-medical applications [3]. For illustrative purposes, we focus attention on a biologically relevant application and material modification.

5.4.1 Supercontinuum Generation in Bio-Media

Consider propagation of ultrashort, intense laser light through homogenous medium containing small quantity of dopant. By way of example, consider a sample of water that contains minute dopant of biological media, such as a protein. Pioneering work on ultrafast supercontinuum generation in water carried out by Alfano and coworkers [26] provided the first evidence of enhancement in supercontinuum generation upon addition of small quantity of inorganic dopants like Zn^{2+} and K^{+}. In more recent work [95], it has been observed that supercontinuum generation is also highly susceptible to changes in concentration of protein like α-amylase. α-amylase is found in human saliva and is regarded as a potential marker of stress; quantification of minutes quantities of this protein in saliva is consequently of importance from the viewpoint of biomedical applications [95].

A typical supercontinuum generation spectra from neat distilled water is shown in Fig. 5.12 and comprises two components, one due to SPM that results in symmetric broadening around the incident laser wavelength, and the other that is asymmetric on the blue side of the spectrum that is generated due to the space–time focusing, self steepening, and plasma formation. Linear chromatic dispersion is sometimes found to play a major role [23] in suppressing supercontinuum in water, in addition to MPI and plasma defocusing. There is a marked dip at 630 nm that is due to an inverse Raman effect [96] in the blue side of spectra. With addition of proteins, it is seen that the symmetric part of the supercontinuum essentially remains unaffected, implying that with addition of proteins to water SPM remains unaffected. The

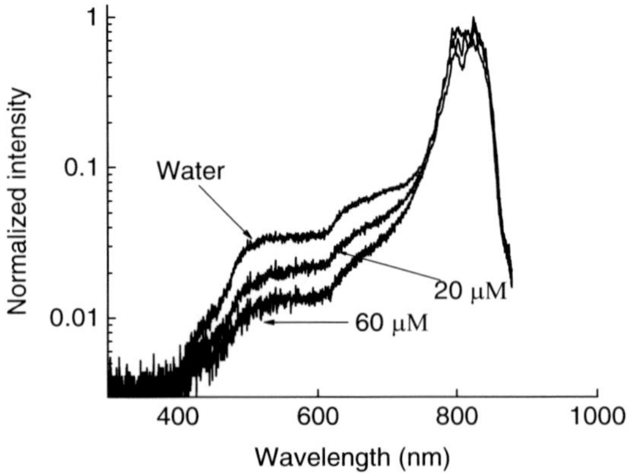

Fig. 5.12. White light spectra obtained from pure water and from water doped with minute quantities of the salivary protein,α-amylase. Note the suppression of the blue-side of the spectrum upon addition of the protein

difference in continuum generation on the addition of α-amylase with concentration changes (20–60 μM) is observed most in the asymmetric part, with the spectral intensity in the blue side showing marked reduction at constant incident laser power [95]. This is also reflected in extinction of the inverse Raman dip. A model has been proposed [95] to rationalize such suppression. As soon as free electrons are generated by MPI, they are rapidly and efficiently scavenged by α-amylase via dissociative attachment. The contribution of electrons to the blue side of the spectrum is reduced as quenching of free electrons takes place, and there is a consequent reduction in the asymmetry of the spectrum. It is noteworthy that supercontinuum generation in water with addition of other molecules such as immunoglobulin, lysine, and arginine does not show significant suppression.

5.4.2 Material Modification

Propagation of pulses of intense light through condensed media also opens exciting opportunities for ultrfast laser-induced bulk modification of transparent materials. There exist a large number of experimental and theoretical works conducted to understand the mechanisms of permanent index modification and laser damage [97–99]. Damage caused by ultrashort pulses is deterministic and is apparently controllable on a microscopic scale [100]. Though material modification produced in transparent materials by ultrashort laser pulses have been related to concepts like micromachining [12, 101], a subject of much contemporary interest is the role of filamentation in permanent index modification. If the index of the medium is permanently modified with an increase in its value without any structural damage, it can act as a waveguide. Davis and coworkers [12] first demonstrated that tightly focused ultrashort laser pulses can indeed write waveguides. During the filamentation with a weakly focused ultrashort laser [102], efficient waveguide structure can be formed. The length of the modified region appears to coincide with the filamentation in the medium and it appears likely that the writing of good waveguides results from a combination of filamentation in weak focusing conditions and melting that results from local heating of the material. Because of the remarkable resistance offered by glass against breakdown and physical damage at 1.5 μm, [103] have realized efficient small-size core optical waveguides in silica based on filamentation process. In contrast, strong focusing conditions lead to the generation of avalanche plasma: a localized explosion can occur and result in void formation [104]. Such voids would not be favorable for the formation of good waveguides [103], yet they may allow the formation of three-dimensional optical memories in the bulk of transparent solids [105]. Recently, Nguyen et al. [106] have written waveguides in fused silica at 800 nm using both filamentation and tight focusing. Channel waveguides written using ultrafast lasers (tightly focused) in erbium-doped phosphate glasses for applications such as integrated amplifiers and lasers operating in the C-band have been demonstrated [107, 108]. Extension of the method to tellurite based glasses has been less successful: though there is one report on writing of short waveguides with positive refractive index change in niobium tellurite glasses [109]. In contrast, only negative refractive index change induced by ultrafast laser irradiation of tellurite glass has been reported, waveguiding action being not possible [110]. We now consider waveguide writing in an erbium-doped tellurite glass modified by phosphate and aluminum addition using 45 fs 806 nm laser [111]. The laser was tightly focused inside transparent tellurite glass using a microscope objective ($25\times$, NA = 0.45), and the glass was translated transversely to the focused beam at three different writing speeds [111]. The channels in tellurite based glass are formed ~500 μm below the surface, with width of 4–8 μm; each channel is 1 cm long. Optical guidance of 1,310 nm light has been observed, indicating a net positive change in refractive index, with losses of $<2\,\mathrm{dB\,cm^{-1}}$.

Acknowledgements

Financial support from the Department of Science and Technology is gratefully acknowledged. We also acknowledge the skills and continuing enthusiasm of colleagues F.A. Rajgara, J.A. Dharmadhikari, H.S. Schroeder, C. Santhosh, and K. Alti.

References

1. S.L. Chin, W. Liu, F. Théberge, Q. Luo, S.A. Hosseini, V. Kandidov, O. Kosareva, N. Aközbek, A. Becker, H. Schroeder, in *Progress in Ultrafast Intense Laser Science III*, ed. by K. Yamanouchi, S.L. Chin, P. Agostini, G. Ferrante (Springer, Berlin, 2008), pp. 243–264.
2. A. Couairon, A. Mysyrowicz, Phy. Rep. **441**, 47 (2007)
3. L. Berg, S. Skupin, R. Nuter, J. Kasparian, J-P. Wolf, Rep. Prog. Phys. **70** 1633 (2007)
4. M. Rodriguez, R. Sauerbrey, H. Wille, L. Wste, T. Fujii, Y.-B. Andr, A. Mysyrowicz, L. Klingbeil, K. Rethmeier, W. Kalkner, J. Kasparin, E. Salmon, J. Yu, J.-P. Wolf, Opt. Lett. **27**, 772 (2002)
5. H. Pépin, D. Comptois, F. Vidal, C.Y. Chien, A. Despaarois, T.W. Johnston, J.C. Kieffer, B.L. Fontaine, F. Martin, F.A.M. Rizk, C. Potvin, P. Couture, H.P. Mercure, A. Bondiou-Clergerie, P. Lalande, I. Gallimberti, Phys. Plasma **8**, 2532 (2001)
6. X.M. Zhao, J.-C. Diels, C.Y. Wang, J.M. Elizondo, IEEE J. Quant. Electron. **QE-31**, 599 (1995)
7. J.F. Gravel, Q. Luo, D. Boudreau, X.P. Tang, S.L. Chin, J. Anal. Chem. **76**, 4799 (2004)

8. J. Kasparin, M. Rodriguez, G. Mjean, J. Yu, E. Salmon, H. Wille, R. Bourayou, S. Frey, Y.-B. Andr, A. Mysyrowicz, R. Saurbrey, J.P. Wolf, L. Wöste, Science **301**, 61 (2003)
9. P. Rairoux, H. Schillinger, S. Niedermeier, M. Rodriguez, F. Ronneberger, R. Saurbrey, B. Stein, D. Waite, C. Wedekind, H. Wille, L. Wöste, C. Ziener, Appl. Phys. B **71**, 573 (2000)
10. H. Kumagi, S.H. Cho, K. Ishikawa, K. Midorikawa, M. Fujimoto, S. Aoshima, Y. Tsuchiya, J. Opt. Soc. Am. B **20**, 597 (2003)
11. E.N. Glenzer, E. Mazur, Appl. Phys. Lett. **71**, 882 (1997)
12. K.M. Davis, K. Miura, N. Sugimoto, K. Hirao, Opt. Lett. **21**, 1729 (1996)
13. L.C. Courrol, R.E. Samad, L. Gomez, I.M. Ranieri, S.L. Baldochi, A. Zanardi de Freitas, N.D. Vieira, Opt. Express **12**, 288 (2004)
14. K. Miura, J. Qiu, H. Inouye, T. Mitsuyu, K. Hirao, Appl. Phys. Lett. **71**, 3329 (1997)
15. H.B. Sun, Y. Xu, S. Juodkazis, K. Sun, M. Watanabe, S. Matsuo, H. Misawa, J. Nishii, Opt. Lett. **26**, 325 (2001)
16. M. Watanabe, H.-B. Sun, S. Juodkazis, T. Takahashi, S. Matsuo, Y. Suzuki, J. Nishi, H. Misawa, Jpn. J. Appl. Phys. **37**, L1527 (1998)
17. E.N. Glenzer, M. Milosavljevic, L. Huang, R.J. Finlay, T.H. Her, J.P. Callan, E. Mazur, Opt. Lett. **21**, 2023 (1996)
18. J.V. Moloney, M. Kolesik, in *Progress in Ultrafast Intense Laser Science II*, ed. by K. Yamanouchi, S.L. Chin, P. Agostini, G. Ferrante (Springer, Berlin, 2008), p. 253
19. W. Liu, S. Petit, A. Becker, N. Aközbek, C.M. Bowden, S.L. Chin, Opt. Commun. **202**, 189 (2002)
20. H. Yang, J. Zhang, L.Z. Zhao, Y.J. Li, H. Teng, Y.T. Li, Z.H. Wang, Z.L. Chen, Z.Y. Wei, J.X. Ma, W. Yu, Z.M. Sheng, Phys. Rev. E **67**, 015401 (2003)
21. A. Brodeur S.L. Chin, J. Opt. Soc. Am. B **16**, 637 (1999)
22. Y. D. Qin, D.L. Wang, S.F. Wang, Q.H. Gong, Chin. Phys. Lett. **18**, 390 (2001)
23. M. Kolesik, G. Katona, J.V. Moloney, E.M. Wright, Phys. Rev. Lett. **91**, 043905 (2003)
24. K. Ishikawa, H. Kumagi, K. Midorikawa, Phys. Rev. E **66**, 056608 (2002)
25. S. Tzortzakis, L. Birge, A. Couairon, M. Franco, B. Prade, A. Mysyrowicz, Phys. Rev. Lett. **86**, 5470 (2001)
26. R.R. Alfano, *The Supercontinuum Laser Source* (Springer, New York, 1989)
27. G.Y. Yang, Y.R. Shen, Opt. Lett. **9**, 510 (1984)
28. M. Maier, W. Kaiser, J.A. Giordmaine, Phys. Rev. **177**, 580 (1969)
29. G.G. Luther, J.V. Moloney, A.C. Newell, E.M. Wright, Opt. Lett. **19**, 862 (1994)
30. P. Chernev, V. Petrov, Opt. Lett. **17**, 172 (1992)
31. J. Kasparian, in *Progress in Ultrafast Intense Laser Science II.*, ed. by K. Yamanouchi, S.L. Chin, P. Agostini, G. Ferrante (Springer, Berlin, 2008), pp. 281–300
32. S. Tzortzakis, L. Sudrie, M. Franco, B. Prade, A. Mysyrowicz, A.Couairon, L. Bergé, Phys. Rev. Lett. **87**, 213902 (2001)
33. W. Liu, O. Kosareva, I.S. Golubtsov, A. Iwasaki, A. Becker, V.P. Kandidov, S.L. Chin, Appl. Phys. B **76**, 215 (2003)
34. A. Dubietis, G. Tamosauskas, I. Diomin, A. Varanavicius, Opt. Lett. **28**, 1269 (2003)

35. J. Philip, C. D'Amico, G. Chériaux, A. Couairon, B. Prade, A. Mysyrowicz, Phys. Rev. Lett. **95**, 163901 (2005)
36. X.W. Chen, Y. Zhu, J. Liu, Y.X. Leng, X.C. Ge, R. Li, Z. Xu, Acta Phys. Sin. **54**, 5178 (2005)
37. R. Li, X. Chen, J. Liu, Y. Zhu, X. Ge, H. Lu, L. Lin, Z. Xu, in *Progress in Ultrafast Intense Laser Science I* ed. K. Yamanouchi, S.L. Chin, P. Agostini, G. Ferrante (Springer, Berlin, 2006), pp. 259–273.
38. G.F. Knoll, *Radiation Detection and Measurement* (Wiley, New York, 1989)
39. A.K. Dharmadhikari, F.A. Rajgara, D. Mathur, Appl. Phys. B **82**, 575 (2006)
40. A.K. Dharmadhikari, F.A. Rajgara, N.C.S. Reddy, A.S. Sandhu, D. Mathur, Opt. Express **12**, 695 (2004)
41. J.H. Marburger, Prog. Quant. Electr. **4**, 35 (1975)
42. M.R. Junnarkar, Opt. Commun. **195**, 273 (2001)
43. G. Mèchain , A. Couairon, M. Franco, B. Prade, A. Mysyrowicz, Phys. Rev. Lett. **93**, 035003 (2004)
44. Z. Jin, J. Zhang, M.H. Xu, X. Lu, Y.T. Li, Z.H. Wang, Z.Y. Wei, X.H. Yuan, W. Yu, Opt. Express **13**, 10424 (2005)
45. W. Liu, F. Theberge, J.F. Daigle, P.T. Simard, S. M. Sarifi, Y. Kamali, H.L. Xu, S.L. Chin, Appl. Phys. B **85**, 55 (2006)
46. G. Fibich, Y. Sivan, Y. Ehrlich, E. Louzon, M. Fraenkel, S. Eisenmann, Y. Katzir, A. Zigler, Opt. Express **14**, 4946 (2006)
47. A.K. Dharmadhikari, K. Alti, J.A. Dharmadhikari, D. Mathur, Phys. Rev. A **76**, 033811-1-5 (2007)
48. A.K. Dharmadhikari, J.A. Dharmadhikari, D. Mathur, App. Phys. B **94**, 259 (2009)
49. I. Golub, Opt. Lett. **15**, 305 (1990)
50. Q. Xing, K.M. Yoo, R.R. Alfano, Appl. Optic **32**, 2087 (1993)
51. C. Conti, S. Trillo, P.D. Trapani, G. Valiulis, A. Piskarskas, O. Jedrkiewicz, J. Trull, Phys. Rev. Lett. **90**, 170406 (2003)
52. S. Skupin, R. Nuter, L. Bergé, Phys. Rev. A **74**, 043813 (2006)
53. A. Braun, G. Korn, X. Liu, D. Du, J. Squier, G. Mourou, Opt. Lett. **20**, 73 (1995)
54. W. Koechner, in *Solid State Laser Engineering* (Springer, Berlin, 1988)
55. H. Schroeder, J. Liu, S.L. Chin, Opt. Express **12**, 4768 (2004)
56. A. Dubietis, G. Tamosauskas, G. Fibich, B. Ilan, Opt. Lett. **29**, 1126 (2004)
57. G. Fibich, S. Eisenmann, B. Ilan, A. Zigler, Opt. Lett. **29**, 1772 (2004)
58. K. Cook, A.K. Kar, R.A. Lamb, Opt. Express **33**, 2025 (2005)
59. G. Heck, J. Sloss, R.J. Levis, Opt. Commun. **259**, 216 (2006)
60. J. Liu, H. Schroeder, S.L. Chin, R. Li, Z. Xu, Appl. Phys. Lett. **87**, 161105 (2005)
61. M. Centurion, Y. Pu, M. Tsang, D. Psaltis, Phys. Rev. A **71**, 063811 (2005)
62. M. Centurion, Y. Pu, D. Psaltis, J. Appl. Phys. **100**, 063104 (2006)
63. A.K. Dharmadhikari, F.A. Rajgara, D. Mathur, H. Schroeder, J. Liu, Opt. Express **13**, 8555 (2005)
64. D. Faccio, M. Porras, A. Dubietis, F. Bragheri, A. Couairon, P. Di Trapani, Phys. Rev. Lett. **96**, 193901 (2006)
65. R.L. Fork, C.V. Shank, C. Hirlimann, R. Yen, W.J. Tomlinson, Opt. Lett. **8**, 1 (1983)
66. P.B. Corkum, C. Rolland, T. Srinivasan-Rao, Phys. Rev. Lett. 57 **18**, 2268 (1986)

67. J.T. Manassah, P.L. Baldeck, R.R. Alfano, Opt. Lett. **13**, 1090 (1988)
68. A. Penzkofer, A. Seilmeier, W. Kaiser, Opt. Commun. **14**, 363 (1975)
69. A.L. Gaeta, Phys. Rev. Lett. **84**, 3582 (2000)
70. N. Aközbek, M. Scalora, C.M. Bowden, S.L. Chin, Opt. Commun. **191**, 353 (2001)
71. X.J. Fang, T. Kobayashi, Appl. Phys. B **77**, 167 (2003)
72. D. Schumacher, Opt. Lett. **27**, 451 (2002)
73. N. Bloembergen, Opt. Commun. **8**, 285 (1973)
74. J.K. Ranka, R.W. Schirmer, A.L. Gaeta, Phys. Rev. Lett. **77**, 3783 (1996)
75. A. Brodeur, S.L. Chin, Phys. Rev. Lett. **80**, 4406 (1998)
76. F. Bragheri, C. Liberale, V. Degiorgio, D. Faccio, A. Matijosius, G. Tamosauskas, A. Varanavicius, P. Di Trapani, Opt. Commun. **256**, 166 (2005)
77. A.K. Dharmadhikari, F.A. Rajgara, D. Mathur, Appl. Phys. B **80**, 61 (2005)
78. A.S. Sandhu, S. Banerjee, D. Goswami, Opt. Commun. **181**, 101 (2000)
79. A.M. Perelomov, V.S. Popov, M.V. Terent'ev, Sov. Phys. JETP **23**, 924 (1966)
80. S. Petit, A. Talebpour, A. Proulx, S.L. Chin, Opt. Commun. **175**, 323 (2000)
81. M. Kolesik, J.V. Moloney, E.M. Wright, Phys. Rev. E **64**, 046607 (2001)
82. G. Fibich, B. Ilan, Phys. Rev. Lett. **89**, 013901 (2002)
83. P. Lambropoulos, Phys. Rev. Lett. **29**, 453 (1972)
84. G.P. Agarwal, in *Nonlinear Fiber Optics* (Academic Press, New York, 2001)
85. K. Midorikawa, H. Kawano, A. Suda, C. Nagura, M. Obara, Appl. Phys. Lett. **80**, 923 (2002)
86. R.R. Alfano, S.L. Shapiro, Phys. Rev. Lett. **24**, 584 (1970)
87. A.K. Dharmadhikari, F.A. Rajgara, D. Mathur, Opt. Lett. **31**, 2184 (2006)
88. D. Mathur, F.A. Rajgara, A.K. Dharmadhikari, J. Phys. Chem. A, **111**, 9399 (2007)
89. S.L. Chin, S. Petit, F. Borne, K. Miyazaki, Jpn. J. Appl. Phys. **38**, L126 (1999)
90. W. Watanabe, K. Itoh, Jpn. J. Appl. Phys. **40**, 592 (2001)
91. K. Cook, A.K. Kar, R.A. Lamb, App. Phys. Lett. **83**, 3861 (2003)
92. C. Corsi, A. Tortora, M. Bellini, Appl. Phys. B **78**, 299 (2004)
93. V. Kandidov, O.G. Kosareva, I.S. Golubtsov, W. Liu, A. Becker, N. Akzöbek, C.M. Bowden, S.L. Chin, Appl. Phys. B **77**, 149 (2003)
94. C.G. Wahlstrm̈, J. Larsson, A. Persson, T. Starczewski, S. Svansberg, P. Salieres, P. Balcou, A. Huillier, Phys. Rev. A **48**, 4709 (1993)
95. C. Santhosh, A.K. Dharmadhikari, K. Alti, J.A. Dharmadhikari, D. Mathur, J. Bio. Med. Opt. **12**, 020510 (2007)
96. W.J. Jones, B.P. Stoicheff, Phys. Rev. Lett. **13**, 657 (1964)
97. B.C. Stuart, M.D. Feit, S. Herman, A.M. Rubenchik, B.W. Shore, M.D. Perry, Phys. Rev. B **53**, 1749 (1996)
98. C.B. Schaffer, A. Brodeur, E. Mazur, Meas. Sci. Technol. **12**, 1784 (2001)
99. S.S. Mao, F. Quéré, S. Guizard, X. Mao, R.E. Russo, G. Petite, P. Martin, Appl. Phys. A **79**, 1695 (2004)
100. A.P. Joglekar, H. Liu, G.J. Spooner, E. Meyhöfer, G. Mourou, A.J. Huni, Appl. Phys. B **77**, 25 (2003)
101. C.B. Schaffer, A. Brodeur, J.F. Garcia, E. Mazur, Opt. Lett. **26**, 93 (2001)
102. L. Sudrie, M. Franco, B. Prade, A. Mysyrowicz, Opt. Commun. **171**, 279 (1999)
103. A. Saliminia, R. Vallée, S.L. Chin, Opt. Commun. **256**, 422 (2005)
104. E.G. Gamaly, S. Juodkazis, K. Nishimura, H. Misawa, B. Luther-Davies, L. Hallo, P. Nicolai, V.T. Tikhonchuk, Phys. Rev. B **73**, 214101 (2006)

105. E.N. Glezer, M. Milosavljevic, L. Huang, R.J. Finlay, T.H. Her, J.P. Callan, E. Mazur, Opt. Lett. **21**, 2023 (1996)
106. N.T. Nguyen, A. Saliminia, S.L. Chin, R. Vallée, Appl. Phys. B **85**, 145 (2006)
107. S. Taccheo, G. D. Valle, R. Osellame, G. Cerullo, N. Chiodo, P. Laporta, O. Svelto, A. Killi, U.M.M. Lederer, D. Kopf, Opt. Lett. **29**, 2626 (2004)
108. R. Osellame, S. Taccheo, G. Cerullo, M. Marangoni, D. Polli, R. Ramponi, P. Laporta, S.D. Silvestri, Electron. Lett. **38**, 964 (2002)
109. Y. Tokuda, M. Saito, M. Takahashi, K. Yamada, W. Watanabe, K. Itoh, T. Yoko, J. Non-Crystal. Solid **326**, 472 (2003)
110. G.C. Righini, I. Banyasz, S. Barneschi, M. Brenci, M. Cremona, D. Ehrt, M. Ferrari, R.M. Montereali, G.N. Conti, S. Pelli, S. Sebastiani, C.Tosello, in *Photonic Materials, Devices, and Applications 2005*, ed. by G. Badenes, D. Abbott, A. Serpenguzel, Proc. SPIE **5840**, pp. 649–657
111. P. Nandi, G. Jose, C. Jayakrishnan, S. Debbarma, K. Chalapathi, K.M. Alti, A.K. Dharmadhikari, J.A. Dharmadhikari, D. Mathur, Opt. Express **14**, 12145 (2006)

6

On Lightning Control Using Lasers

Jérôme Kasparian and Jean-Pierre Wolf

Summary. After a quick review of historic lightning research, we discuss developments aiming at triggering and guiding lightning using lasers. We recall the limitations inherent to high-energy, nanosecond pulses and discuss the new perspective introduced by the advent of ultrashort, high-power laser pulses and the long ionized filaments they induce in air. After describing a recent field campaign where the triggering of electric activity in thunderclouds by ultrashort lasers was demonstrated for the first time, we discuss perspectives for improvements of both the sequence of laser pulses and the experimental configuration.

6.1 Introduction

Lightning, an unpredictable and frightening phenomenon, has been associated with God's anger up to the Middle Ages [1]. Research for natural interpretations of lightning began in the seventeenth Century when René Descartes attributed it to collisions between clouds. However, lightning studies really developed in the eighteenth Century, in parallel with the development of knowledge about static electricity. In 1752, in Marly-la-Ville, near Paris, Thomas François Dalibard observed sparks between a conducting stick pointed towards a thundercloud and isolated from ground on one side and a grounded tip on the other side. This experiment, which had been proposed by Benjamin Franklin (1706–1790), confirmed that lightning is an electrical phenomenon [2]. On the next year, by grounding the conducting stick, Franklin invented the lightning rod. In the twentieth Century, the advent of photography and of the cathodic ray tube oscilloscope allowed a precise characterization of lightning. Those techniques allowed to observe the trajectory of lightning strikes, and to measure the typical involved voltage ($\sim$100 MV) and intensity ($\sim$30 kA). Today's instruments and vectors, such as balloon, aircrafts, or satellites, provide an even more precise view of the processes at play in the physics of lightning.

6.2 The Lightning Strike

The preliminary step for a lightning strike is the electric loading of clouds. Charges are generated by collisions between ice and/or water particles, and separated by vertical winds within clouds at speeds up to $20\,\mathrm{m\,s^{-1}}$. Positively charged ice particles accumulate at the top of the cloud, while the bottom of the cloud is negatively charged. This charge separation generates an electric field of up to 10–15 $\mathrm{kV\,m^{-1}}$ at ground level, and $50\,\mathrm{kV\,m^{-1}}$ some hundreds of meters above ground.

The deformation of the charged water droplets of the cloud in this electric field forms ellipsoids which further increase the electric field and initiate corona discharges at their tip. These corona discharges develop into streamers, which connect and form a leader (Fig. 6.1a). This ionized channel progresses by steps of some tens of meters, with rest times of 50–100 μs between two steps (Fig. 6.1b). During its development, the leader splits into several branches. When one branch reaches close to ground, an upward leader is generated from an elevated point such as a tree, a building, or a mountain ridge. The connection of the two leaders allows the flow of thousands to tens of thousands of Amperes from the ground to the cloud: This high current flow, or *return stroke* (Fig. 6.1c), constitutes the visible part of the lightning strike and lasts from a fraction of a second to a few seconds, including re-illuminations when the same channel is used by subsequent current pulses.

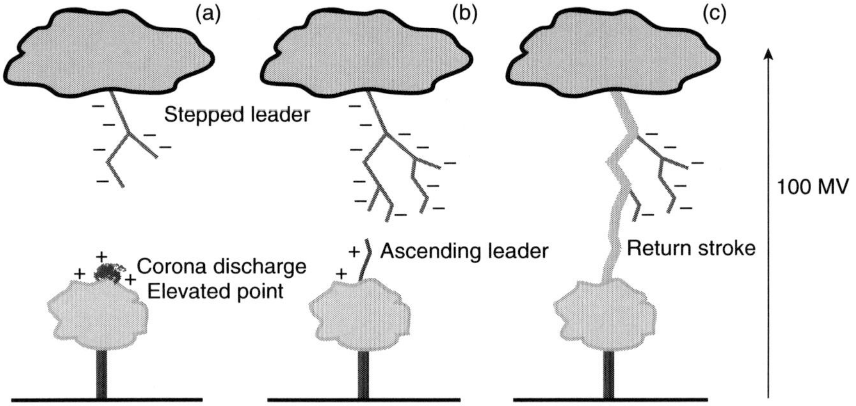

Fig. 6.1. Mechanism of lightning initiation: (**a**) stepped leader formation; (**b**) initiation of an ascending leader; (**c**) return stroke

6.3 Attempts to Trigger Lightning Using High-Energy Lasers

The electron avalanches of the streamer-leader mechanism typically need 10 m to develop. Such distance is generally not available in laboratory experiments because of the limitations in the available voltages and installation sizes. Field experiments on real-scale lightning are therefore necessary, but the random character of lightning limits its availability at the experiment location and prevents any synchronization of the instrumentation with the lightning strike itself. Therefore, several groups [3,4] developed techniques to trigger on-demand lightning with a rocket pulling a thin metallic wire, which initiates a lightning strike and guides it to the ground. Lightning strikes can even be triggered by short conducting wire sections, which perturb less the lightning mechanism than a continuous conducting wire does. However, the number of rockets available during one thunderstorm is limited to 5–10 in practice. Also, the rocket has to be launched with the right timing relative to the increase of the electric field in the cloud. Therefore, a continuously running lighting triggering system would be suitable.

Lasers have been identified as a candidate in this purpose as early as in the 1960s, as reviewed in [5]. Long laser sparks were produced using neodymium or CO_2 lasers, with energies up to several kilo-joules [6] and a typical pulse duration of 50 ns [7]. With such high energy, the channel can be heated to several thousands of degrees, which reduces the gas density by typically one order of magnitude. As the rate of collisional ionization by electrons is inversely proportional to the gas density, such heating is highly favorable to the air ionization. Moreover, above a temperature of 4,000 K, the associative ionization of radicals created by the laser pulse $N + O \rightarrow NO^+ + e^-$ contributes significantly to the ionization process, independently from the electric field. Heating also mostly suppresses the electron losses due to their attachment and recombination. However, by efficiently ionizing the air, the leading edge of the pulse generates a dense plasma, which is opaque to the trailing edge of the pulse. A large fraction of the pulse energy is therefore lost for further propagation, preventing the generation of a connected plasma channel beyond a length of a few meters. Even a geometrical focus tens to hundreds of meters away cannot generate connected plasma channels, but rather a series of localized plasma sparks.

In spite of these limitations, field experiments using high-energy lasers [8] were carried out to intercept lightning with lasers on the shore of the Sea of Japan. The experiments were conducted in a period of intense winter low-cloud thunderstorms, with electric fields up to $10\,\mathrm{kV\,m^{-1}}$. A first CO_2 laser delivering 1 kJ was focused on a dielectric target at the top of a 50 m high tower constructed on a 200 m high hill. A second one was aimed near to the dielectric to form a 2 m long plasma spark. A third, ultraviolet laser was producing a weakly ionized plasma channel, slightly offset from the tower, to direct the leader to the cloud. The natural initiation of cloud discharges was considered as the precursor of the descending lightning strikes. Therefore, the lasers were triggered on the detection of such discharges. The authors reported two successful attempts, although the statistical significance of their result is not clear.

6.4 Control of High-Voltage Discharges Using Ultrashort Lasers

In contrast to longer (nanosecond) pulses, ultrashort laser pulses in air can form plasma channels [9–13] with a length up to 100 m [14] at a distance of several kilometres [15]. This process named *filamentation* stems from a dynamic balance between the Kerr self-focusing on one side and the defocusing effect of the free electrons of the plasma generated at the nonlinear focus on the other side (Fig. 6.2). Moreover, steering the beam allows to aim the ionized plasma channel at the most active part of a thundercloud. Furthermore, we recently showed that filaments can propagate almost unperturbed in adverse conditions such as rain [16], fog [17], turbulence [18, 19], or reduced pressure [16], which makes them highly suitable for atmospheric applications.

The first demonstration of the ability of ultrashort laser pulses to trigger high-voltage discharges have been performed using focused infrared [20, 21] or ultraviolet [22] lasers. Later, spectacular experiments with the *Teramobile* laser [23] installed in a high-voltage facility showed that laser filaments can trigger and guide 1.8 MV discharges over up to 4.5 m. The breakdown voltage is typically reduced by 30% [24,25]. Moreover, triggered discharges are guided along the laser beam (Fig. 6.3). Partly guided discharges also occur in some configurations, providing information about the mechanism of the initiation and propagation of laser-triggered discharges [26]. Even artificial rain does not prevent the laser filaments to trigger discharges [27]. Current work focuses on the possibility to extend the plasma lifetime using auxiliary lasers in order to

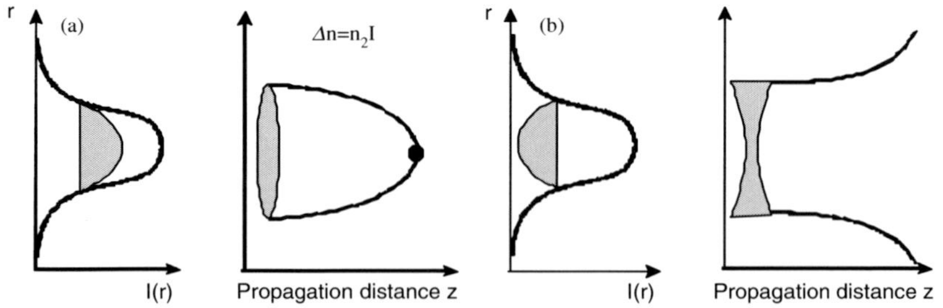

Fig. 6.2. Mechanism of filamentation. (**a**) Kerr effect (self-focusing); (**b**) defocusing by the plasma

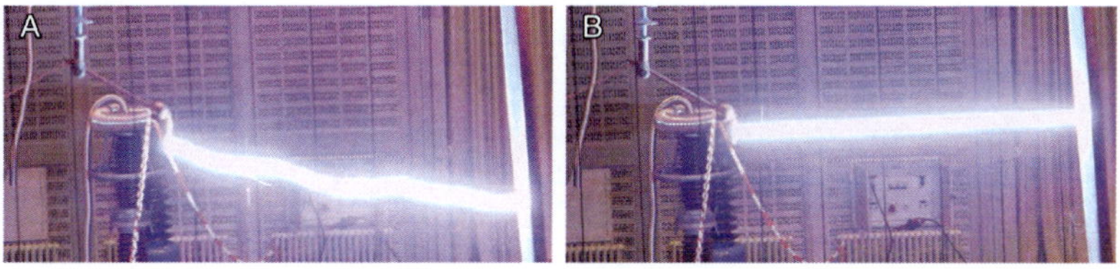

Fig. 6.3. Laser control of high-voltage discharges. (**a**) Free discharge over 3 m, without laser filaments. Note the erratic path. (**b**) Straight discharge guided along laser filaments [25]

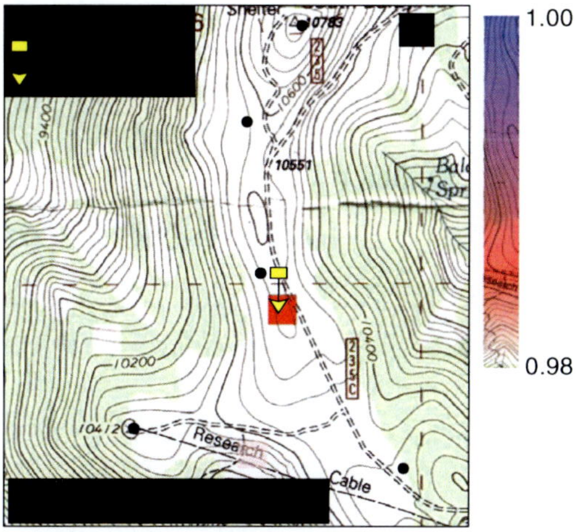

Fig. 6.4. Statistical confidence level associated with the electric events synchronized with the laser pulses during a thunderstorm. The color scale is transparent below 98% (i.e., for error risks above 2%), leaving the topographical background uncovered. *Arrowhead*: location of the laser-induced plasma channel; *Arrow tail*: laser emitter. Topographic background courtesy of US Geological Survey [31]

increase the possible guiding length and improve the scalability to atmospheric scales. This approach relies on reheating and photodetaching electrons of the plasma channel by subsequent pulses, either in the nanosecond [28] or in the femtosecond regime [29]. Although a high power of the subsequent laser pulse is generally considered necessary for efficiently photodetaching electrons from O_2^- ions in the plasma, it was recently demonstrated that a YAG laser pulse of moderate energy (sub-Joule) at 532 nm significantly increases the triggering effect of an infrared femtosecond laser near to the voltage threshold [30]. This effect was interpreted as resulting from a positive retroaction loop where Joule heating of the plasma channel enhances photodetachment, while the resulting higher electron density boosts the Joule effect.

6.5 Field Experiments Using Femtosecond Laser Filamentation

Following the demonstration of both the capability of laser filaments to propagate in realistic atmospheric conditions and their ability to trigger high-voltage electric discharges, the effect of femtosecond plasma channels of moderate energy on thunderclouds was investigated during a field campaign [31] at the Langmuir Laboratory (New Mexico, USA, see Fig. 6.4), which provides a fully equipped facility with high lightning occurrence.

The *Teramobile* femtosecond–terawatt laser [23] was fired from the ridge of South Baldy Peak, 3,209 m above sea level, at a repetition rate $f = 10$ Hz. It emitted a collimated (unfocused) beam, leaning southwards 70° above horizontal. The laser beam diameter was 3 cm, and the energy per pulse was 270 mJ at a central wavelength of 800 nm. The laser pulses, of 150 fs initial duration, were negatively chirped to 600 fs to generate multiple filamentation with significant ionization over a typical length of 100 m, a few hundreds of meters above the ground.

Two thunderstorms were investigated, focusing on the overall 250 s (i.e., 2,500 laser pulses) when the electric field would have been sufficient to trigger lightning using rockets. Five Lightning Mapping Array (LMA) [32] receivers located within 1 km distance from the laser (see Fig. 6.4) detected radiofrequency (RF) pulses generated by the electric activity in the atmosphere, with a time accuracy of 40 ns. The comparison of the times of arrival of the RF pulses on the detectors yielded the location and time stamping of their origin. Sets of RF sources with delay mismatches corresponding to distances up to 100 m were included in the analysis to account for events distributed along the laser beam, that is, over a length of up to 100 m.

Because the laser is fired independently from the electric activity in the clouds, and because natural electric activity in clouds is a random process, temporal correlations between the electric events detected by the LMA and the laser operation were used as an evidence for an effect of the laser. At each location, the events synchronized (within 2 ms) with the laser have been identified. The probability α_{sync} that these events may have been obtained by chance among random events, rather than being due to an effect of the laser, was estimated. α_{sync} can be understood as the risk of error when concluding that the observed pulses are related with the laser pulses. This corresponding confidence level is therefore $1 - \alpha_{\text{sync}}$.

Figure 6.4 displays the result of this statistical analysis for one single thunderstorm. At the location of the laser filaments (arrow head), 43% (3 out of 7) of the pulses are synchronized with the laser repetition rate, corresponding to a high statistical significance ($1 - \alpha_{sync} = 0.987$). The delay mismatch between the RF pulses detected on the different LMA detectors correspond to some tens of meters, typical of spatially spread events, such as a series of simultaneous corona discharges at the tips of individual filaments within the bundle of multiple filamentation.

Similar results have been obtained during the second thunderstorm [31]. In contrast, no effect was observed when the electric field was low or negative. Therefore, these results suggest that a small fraction (0.24% of the laser pulses, i.e., ~1 event/minute) of the plasma filaments have initiated electric events in a strong positive (upward pointing) electric field [31].

Plasma filaments generated in the atmosphere by ultrashort laser pulses therefore appear to be able to trigger electric events in thunderclouds under high positive electric field. This result constitutes a first step towards laser-controlled lightning.

6.6 Optimization of the Filament Effect in Thunderstorms

6.6.1 Optimization of the Plasma Density and Lifetime

The very limited effect of the plasma channel left behind by the filaments, which did not trigger lightning strikes to the ground, appears to be due to their low electron density as well as their limited lifetime of ~1 μs [33,34]. This duration corresponds to an effective length on the meter-scale for a leader propagating at a speed of a few $10^6\,\mathrm{m\,s^{-1}}$ [24]. As wires of a few tens of meters are sufficient for triggering lightning using rockets, the plasma lifetime has to be enhanced up to several tens of micro second. This could be achieved by a subsequent nanosecond, multi-Joule laser pulse [35, 36]. Even a frequency-doubled YAG pulse of moderate energy (200 mJ) is sufficient to improve the ability of femtosecond filaments to trigger high-voltage discharges on the meter-scale [37]. Sequences of ultrashort pulses at 800 nm have also been proposed to improve the plasma lifetime [38, 39]. The definition of the optimal pulse sequence is still an open question, which will require throughout modeling of the temporal evolution of the plasma channel created by the laser in an electric field [40]. Other available parameters include the pulse energy and beam profile. While the latter is difficult to control over long distances in the highly perturbed propagation conditions encountered in thunderstorms, a higher pulse energy could yield more filaments over longer distances [41], thus possibly improving the effect of the laser on the electric field.

6.6.2 Mechanism of the Laser Filament Action in Thunderclouds

The emission geometry can also be optimized. Such optimization first requires a clarification of the mechanism of the expected laser action on the thundercloud.

Two mechanisms can be considered for the triggering of lightning by laser filaments. In a first scenario, the free electrons constitute a seed from which the filament, in the electric field, will turn into an ascending leader, and then into an ascending lightning strike. This scenario may appear realistic at first, as it requires an initial electron density of only $5 \times 10^{11}\,\mathrm{cm}^{-3}$ [35], three to four orders of magnitude below the typical plasma densities in filaments. Also, the required filament length is that of a single step of a leader, that is, a few tens of meters only. However, such filament-to-leader conversion process implies an electron avalanche, and therefore requires a local electric field of at least $26\,\mathrm{kV\,m}^{-1}$. Such field can only be achieved if the filament starts from the top of a high tower (typically beyond 200 m). [1] Moreover, even at high towers, ascending lighting accounts for only 5–10% of the lightning activity, which drastically restricts the number of potential events. Another limitation of this process is in the small number of electrons available in a filament, due to its reduced volume. A typical filament can be modeled as a cylinder of 30 μm diameter and 50 m length, providing a volume of $0.04\,\mathrm{cm}^3$ only. In spite of its electron density of $10^{15}\,\mathrm{cm}^{-3}$, it therefore contains only 4×10^{13} electrons. In comparison, the electron density in the corona discharge at the tip of a metallic rod is only 10^{14} electrons cm^{-3} [42], but its volume is orders of magnitude larger. Therefore, a filament cannot be expected to release more electrons in the atmosphere than a passive lightning rod does.

The following discussion will therefore focus on a second scenario, where the filament is considered as a passive conductor. This approach is consistent with the interpretation of the laser-triggered discharges in the laboratory as ohmic bridging [24]. The conducting filament suddenly emerges from the space charge accumulated around an elevated point such as the top of a tower, or near to the ground in a tower-free configuration. The sudden emergence of the filament acts like a rocket with a sufficient speed to precede the build-up of a space charge around its tip and avoid the screening of the electric field around itself.

The action of a plasma channel as a conductor is governed by the loading time of the corresponding RC circuit. The system formed by the ground, the filament, and the cloud can be considered as a serial circuit including two tip-plane capacitors and one resistor. The two capacitors account for the gaps between the filament and the cloud, as well as between the filament and ground, each with a capacity $C = 4\pi\varepsilon_0 rS/(S + d \cdot r) \approx 4\pi\varepsilon_0 r$, where S is the surface of the bottom of the cloud, $r = 100\,\mu\mathrm{m}$ is the radius of curvature of the " tip " at the end of the filament, $d = 100\,\mathrm{m}$ is the distance between the filament end and the cloud (respectively, the ground). The filament itself is modeled as a resistor $R = l \cdot r$, $l = 50\,\mathrm{m}$ being the typical filament length and $r = 1\,\mathrm{M}\Omega\,\mathrm{m}^{-1}$ its resistivity [24]. Under these assumptions, the loading time of the filament is $t = R \cdot C/2 = 10^{-7}\,\mathrm{s}$. While this duration is longer than the picosecond timescale of electron attachment, it is much shorter than the microsecond lifetime of the plasma channel. We therefore neglect the transient and dynamical effects and handle the filament as a passive conductor inserted in the beam. Note that considering a bundle of N filaments yields $R = N \cdot l \cdot \mathrm{r}$

and $C = 4\pi\varepsilon_0 NrS/(S + N \cdot d \cdot r) \approx 4\pi\varepsilon_0 Nr$, so that t is unaffected by the number of filaments in the bundle.

As the filaments are considered as passive, the triggering of lighting depends on the ability of the filaments to intercept a descending leader. In the context of lightning protection engineering, it is common to consider that an object has an influence on leaders in its vicinity, within a distance $r = 10 \times I^{2/3}$, where r is expressed in meters and I the intensity (in kA) of the lightning strike that will occur. A typical value of $I = 30$ kA, yields $r = 96$ m. We consider that, once a filament interacts with the descending leader and generates an upward positive leader, the current increases in the plasma channel and stabilizes it [43]. Within this assumption, the filament intercepts all descending leaders that are within its area of influence r at the time of the laser pulse.

6.6.3 Influence of the Geometric Configuration

To illustrate the impact of several configuration parameters with and without tower, we performed phenomenological Monte-Carlo simulations of a descending leader above a flat ground. In each run, a stepped leader is launched from a random location in a cloud 2,000 m above ground, and propagates in 10 m steps. The direction of each step is randomly chosen over 360° in azimuth and from 0 to −90° in elevation. When the leader reaches a distance smaller than r from the tower, filament, or ground, the lightning strike is considered to hit the corresponding element. The simulation is conducted over a $3 \times 3\,\text{km}^2$ area. In this simple phenomenological model, the filament is modeled as a permanently available conductor. We checked that the simulation results are almost insensitive to the topography, provided reasonable landscape shapes are considered.

We investigated several geometries. In a first geometry (Fig. 6.5a), the beam is aimed at the top of a 200 m high tower, extending its length vertically. Such a configuration is comparable to that of Uchida et al. [8]. According to our simple model, a 200 m long filament on top of the tower only increases the number of lightning strikes on it by 10–20%. This increase hardly exceeds the 13% standard deviation for the number of strikes on the tower alone. Therefore, this first configuration is very unfavorable for field experiments because it will not allow to observe a statistically significant increase of the number of events for reasonable filament lengths.

In a second geometry (Fig. 6.5b), the filament is launched vertically parallel to the tower, competing with it to attract lightning. If we consider the competition between a 200 m high tower and a filament located 30 m apart from it, the main observable assessing for an effect of the laser would be the depletion of the number of lightning strikes on the tower as a function of the filament height. The depletion becomes statistically significant (2 standard deviations, i.e., 26%) if the filament reaches a height of 175 m, which in principle can be practically achieved. However, the poor duty cycle of the filament

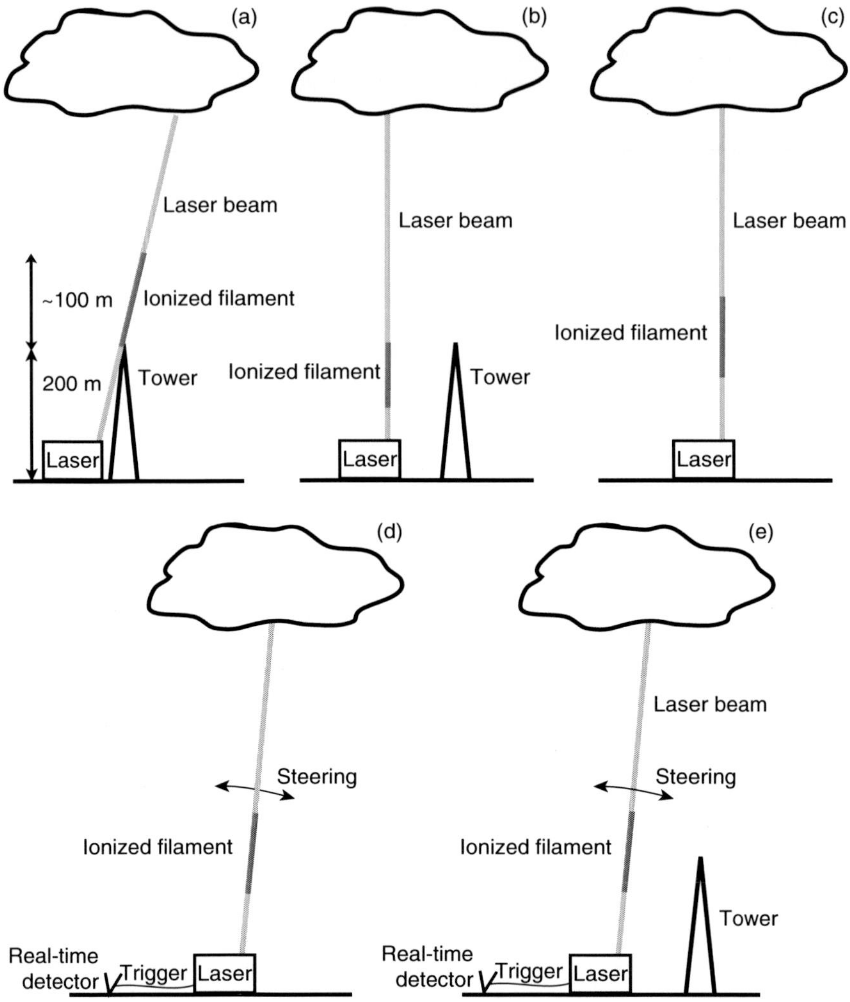

Fig. 6.5. Schematic of the considered configurations. (**a**) Filament extending a tower; (**b**) Filament competing with a tower; (**c**) filament alone; (**d**) Filament with real-time triggering and aiming; (**e**) Filament with real-time triggering and aiming, competing with a tower

has to be taken into account. A descending leader needs a time $t_{\text{transit}}{\sim}100\,\mu\text{s}$ to go across the region of influence of the filament, with $r = 100\,\text{m}$, at a typical propagation speed of $1\,\text{m}/\mu\text{s}$. [24] At 10 Hz ($T = 100\,\text{ms}$), this corresponds to a duty cycle of 1/1000 which jeopardises the hope to observe a statistically significant effect within a finite time.

The third geometry was that of the *Teramobile* team during their field campaign in New Mexico [31]. The filament is launched independently from any tower (Fig. 6.5c). Again, very few descending leaders can be intercepted, independently on the filament altitude and length, especially when taking the duty cycle into account. Considering the occurrence of a few hundreds to thousands of natural descending leaders over the experimental area during a 3-month campaign, typical for a spot of high lightning occurrence, a few events at most can be expected. This could explain the very reduced number of events actually observed during the *Teramobile* field campaign [31].

However, the low efficiency of the above-described strategies may be overcome by taking benefit of one of the most prominent advantages of the laser technique, namely its flexibility. In a fourth configuration, we assume that a real-time detector, as used by Uchida et al. [8] is implemented to detect and locate the advent of descending leaders so as to trigger the laser and aim it at their origin, with or without a competing tower (Figs. 6.5d, e). This approach improves the duty cycle by several orders of magnitude, as all laser shots are able to intercept a descending leader. Moreover, the steering capability drastically extends the volume of influence of the laser. Therefore, it increases the number of accessible descending leaders at least by a factor of 10, even for conservative filament heights of a few hundreds of meters. Intercepting several tens to a few hundreds of shots in one single campaign therefore appears accessible. This number of events would be sufficient to achieve the statistical significance required for such a campaign to be conclusive.

Although phenomenological and simplistic, these simulations show that the efficiency of laser triggering and guiding may be improved by orders of magnitude, provided the laser shots are triggered by electric activity in thunderclouds and kept oriented towards the most active regions of the clouds.

6.7 Conclusion

The 50-years old history of laser-assisted lightning has brought a wealth of laboratory development as well as field work, with mitigated success. While the potential of high-energy, nanosecond lasers appears to be limited, recent field results with the *Teramobile* ultrashort laser opened new prospects to reach the goal of triggering lightning with lasers.

Such an achievement will require to optimize the plasma channels generated during the filamentation of laser pulses. Emitting pulse sequences and increasing the pulse energy shall improve the electron density and the plasma lifetime. In parallel, optimized triggering and beam steering schemes, driven by real-time detectors locating the initial stages of descending stepped leaders, will greatly improve both the duty cycle and effective volume of influence of laser filaments. It should therefore increase by orders of magnitude the number of available events during field campaigns. This development is still in process. However, good hopes can be raised that laser triggering of lightning could be achieved in a near future.

Acknowledgements

This work has been performed in the framework of the *Teramobile* collaboration (www.teramobile.org), funded jointly by the CNRS, DFG, French and German ministries of Foreign affairs, Agence Nationale de la Recherche (ANR, grant # NT05–1_43175), Fonds national suisse de la recherche scientifique (FNS, grants #200021–111688/1 and 200021–116198/1), and the Swiss Secrétariat d'État à l'Éducation et à la Recherche (COST P18 action C06.0114). We acknowledge all our collaborators within the *Teramobile* project for common work and constructive discussions.

References

1. C. Gary, *La foudre. Nature, histoire, risques et protection* (Dunod, Paris, 2004)
2. E.P. Krider, *Benjamin Franklin and the First Lightning Conductors*, Proceedings of International Commission on History of Meteorology 1 1–13, 2004
3. R. Fieux, C. Gary, P. Hubert, Nature **257**, 212–214 (1975)
4. V.A. Rakov, M.A. Uman, K.J. Rambo, Atmos. Res. **76**, 503–517 (2005)
5. E.M. Bazelyan, Y.P. Raizer, Phys. Usp. **43**, 701 (2000)
6. V. Apollonov, et al., AHPLA'99 Paper 3886–34
7. M. Miki, Y. Aihara, T. Shindo, J. Phys. D Appl. Phys. **26**, 1244–1252 (1993)
8. S. Uchida, Y. Shimada, H. Yasuda, S. Motokoshi, C. Yamanaka, T. Yamanaka, Z. Kawasaki, K. Tsubakimoto, J. Opt. Technol. **66**, 199–202 (1999)
9. A. Braun, G. Korn, X. Liu, D. Du, J. Squier, G. Mourou, Opt. Lett. **20**, 73–75 (1995)
10. S.L. Chin, S.A. Hosseini, W. Liu, Q. Luo, F. Theberge, N. Akozbek, A. Becker, V. P. Kandidov, O.G. Kosareva, H. Schroeder, Can. J. Phys. **83**, 863–905 (2005)
11. L. Bergé, S. Skupin, R. Nuter, J. Kasparian, J.-P. Wolf, Rep. Prog. Phys. **70**, 1633–1713 (2007)
12. A. Couairon, A. Mysyrowicz, Phys. Rep. **441**, 47–189 (2007)
13. J. Kasparian, J.-P. Wolf, Opt. Express **16**, 466–493 (2008); M. Rodriguez, R. Bourayou, G. Méjean, J. Kasparian, J. Yu, E. Salmon, A. Scholz, B. Stecklum, J. Eislöffel, U. Laux, A.P. Hatzes, R. Sauerbrey, L. Wöste, J.-P. Wolf, Phys. Rev. E **69**, 036607 (2004)
14. B. La Fontaine, F. Vidal, Z. Jiang, C.Y. Chien, D. Comtois, A. Desparois, T.W. Johnson, J.-C. Kieffer, H. Pépin, Phys. plasma **6**, 1615–1621 (1999)
15. M. Rodriguez, R. Bourayou, G. Méjean, J. Kasparian, J. Yu, E. Salmon, A. Scholz, B. Stecklum, J. Eislöffel, U. Laux, A. P. Hatzes, R. Sauerbrey, L. Wöste, J.-P. Wolf, Phys. Rev. E **69**, 036607 (2004)
16. G. Méchain, G. Méjean, R. Ackermann, P. Rohwetter, Y.-B. André, J. Kasparian, B. Prade, K. Stelmaszczyk, J. Yu, E. Salmon, W. Winn, L.A.V. Schlie, A. Mysyrowicz, R. Sauerbrey, L. Wöste, J.-P. Wolf, Appl. Phys. B **80**, 785–789 (2005)
17. G. Méjean, J. Kasparian, J. Yu, E. Salmon, S. Frey, J.-P. Wolf, S. Skupin, A. Vinçotte, R. Nuter, S. Champeaux, L. Bergé, Phys. Rev. E **72**, 026611 (2005)

18. R. Ackermann, G. Méjean, J. Kasparian, J. Yu, E. Salmon, J.-P. Wolf, Opt. Lett. **31**, 86–88 (2006)
19. R. Salamé, N. Lascoux, E. Salmon, J. Kasparian, J. P. Wolf, Appl. Phys. Lett. **91**, 171106 (2007)
20. H. Pépin, D. Comtois, F. Vidal, C.Y. Chien, A. Desparois, T.W. Johnston, J.C. Kieffer, B. La Fontaine, F. Martin, F.A.M. Rizk, C. Potvin, P. Couture, H.P. Mercure, A. Bondiou-Clergerie, P. Lalande, I. Gallimberti, Phys. Plasma **8**, 2532–2539 (2001)
21. D. Comtois, C.Y. Chien, A. Desparois, F. Gérin, G. Jarry, T.W. Johnston, J.C. Kieffer, B. La Fontaine, F. Martin, R. Mawassi, H. Pépin, F.A.M. Rizk, F. Vidal, P. Couture, H.P. Mercure, C. Potvin, A. Bondiou-Clergerie, I. Gallimberti, "Appl. Phys. Lett. **76**, 819–821 (2000)
22. P. Rambo, J. Schwartz, J.-C. Diels, J. Opt. A Pure Appl. Opt. **3**, 146–158 (2001)
23. H. Wille, M. Rodriguez, J. Kasparian, D. Mondelain, J. Yu, A. Mysyrowicz, R. Sauerbrey, J.-P. Wolf, L. Wöste, Eur. Phys. J. Appl. Phys. **20**, 183–190 (2002)
24. M. Rodriguez, R. Sauerbrey, H. Wille, L. Wöste, T. Fujii, Y.-B. André, A. Mysyrowicz, L. Klingbeil, K. Rethmeier, W. Kalkner, J. Kasparian, E. Salmon, J. Yu, J.-P. Wolf, Opt. Lett. **27**, 772–774 (2002)
25. J. Kasparian, M. Rodriguez, G. Méjean, J. Yu, E. Salmon, H. Wille, R. Bourayou, S. Frey, Y.-B. André, A. Mysyrowicz, R. Sauerbrey, J.-P. Wolf, L. Wöste, Science **301**, 61–64 (2003)
26. R. Ackermann, G. Méchain, G. Méjean, R. Bourayou, M. Rodriguez, K. Stelmaszczyk, J. Kasparian, J. Yu, E. Salmon, S. Tzortzakis, Y.-B. André, J.-F. Bourrillon, L. Tamin, J.-P. Cascelli, C. Campo, C. Davoise, A. Mysyrowicz, R. Sauerbrey, L. Wöste, J.-P. Wolf, Appl. Phys. B. **82**, 561–566 (2006)
27. R. Ackermann, K. Stelmaszczyk, P. Rohwetter, G. Méjean, E. Salmon, J. Yu, J. Kasparian, G. Méchain, V. Bergmann, S. Schaper, B. Weise, T. Kumm, K. Rethmeier, W. Kalkner, J.P. Wolf, L. Wöste, Appl. Phys. Lett. **85**, 5781–5783 (2004)
28. P. Rambo, J. Schwartz, J.-C. Diels, J. Opt. A Pure Appl. Opt. **3**, 146–158 (2001)
29. Z. Hao, J. Zhang, Y.T. Li, X. Lu, X.H. Yuan, Z.Y. Zheng, Z.H. Wang, W.J. Ling, Z.Y. Wei, Appl. Phys. B **80**, 627–630 (2005)
30. G. Méjean, R. Ackermann, J. Kasparian, E. Salmon, J. Yu, J.-P. Wolf, K. Rethmeier, W. Kalkner, P. Rohwetter, K. Stelmaszczyk, L. Wöste, Appl. Phys. Lett. **88**, 021101 (2006)
31. J. Kasparian, R. Ackermann, Y.-B. André, G. Méchain, G. Méjean, B. Prade, P. Rohwetter, E. Salmon, K. Stelmaszczyk, J. Yu, A. Mysyrowicz, R. Sauerbrey, L. Wöste, and J.-P. Wolf, Journal of the European Optical Society: Rapid Publications. 3, 08035 (2008)
32. W. Rison, R.J. Thomas, P.R. Krehbiel, T. Hamlin, J. Harlin, Geophys. Res. Lett. **26**, 3573–3576 (1999)
33. B. La Fontaine, F. Vidal, D. Comtois, C.-Y. Chien, A. Deparois, T.W. Johnston, J.-C. Kieffer, H. P. Mercure, H. Pépin, F.A.M. Rizk, IEEE Trans. Plasma Sci. **27**, 688–700 (1999)
34. S. Tzortzakis, B. Prade, M. Franco, A. Mysyrowicz, Opt. Commun. **181**, 123–127 (2000)
35. X.M. Zhao, J.-C. Diels, C. Y. Wang, J. M. Elizondo, IEEE J. Quant. Electron. **31**, 599–612 (1995)

36. P. Rambo, J. Schwarz, J.-C. Diels, J. Opt. A Pure Appl. Opt. **3**, 146–158 (2001)
37. G. Méjean, R. Ackermann, J. Kasparian, E. Salmon, J. Yu, J.-P. Wolf, K. Rethmeier, W. Kalkner, P. Rohwetter, K. Stelmaszczyk, L. Wöste, Appl. Phys. Lett. **88**, 021101 (2006)
38. H. Yang, J. Zhang, Y. Li, J. Zhang, Y. Li, Z. Chen, H. Teng, Z. Wei, Z. Sheng, Phys. Rev. E **66**, 016406 (2002)
39. Z. Hao, J. Zhang, Y.T. Li, X. Lu, X.H. Yuan, Z.Y. Zheng, Z.H. Wang, W.J. Ling, Z.Y. Wei, Appl. Phys. B **80**, 627–630 (2005)
40. T.B. Petrova, H.D. Ladouceur, A.P. Baronavski, Phys. Rev. E **76**, 066405 (2007)
41. P. Béjot, L. Bonacina, J. Extermann, M. Moret, J.P. Wolf, R. Ackermann, N. Lascoux, R. Salamé, E. Salmon, J. Kasparian, L. Bergé, S. Champeaux, C. Guet, N. Blanchot, O. Bonville, A. Boscheron, P. Canal, M. Castaldi, O. Hartmann, C. Lepage, L. Marmande, E. Mazataud, G. Mennerat, L. Patissou, V. Prevot, D. Raffestin, J. Ribolzi, Appl. Phys. Lett. **90**, 151106 (2007)
42. F. Grangé, N. Soulem, J.F. Loiseau, N. Spyrou, J. Phys. D. Appl. Phys. **28** 1619–1629 (1995)
43. T.B. Petrova, H.D. Ladouceur, A.P. Baronavski, Phys. Plasma **15**, 053501 (2008)

7

Advances in X-Ray Studies of Ultraintense Laser-Plasma Interactions

Leonida A. Gizzi

Summary. This Chapter focuses on the investigation of fast electron transport studies in solids irradiated at relativistic laser intensities. Experimental techniques based upon space-resolved spectroscopy are presented in view of their application to both ultrashort Kα X-ray sources and fast ignition studies. Spectroscopy based upon single-photon detection is unveiled as a complementary diagnostic technique, alternative to well established techniques based upon bent crystals. Application of this technique to the study of X-ray fluorescence emission from fast electron propagation in multilayer targets is reported and explored as an example case.

7.1 Introduction

Laser plasma interactions are well known for their X-ray emitting properties [1, 2], from the nanosecond down to the sub-femtosecond temporal domain. Because of their micrometric source size and their high peak brilliance, X-ray sources based on laser–matter interactions are of a great interest for fundamental studies and for a variety of applications. The development of chirped pulse amplification (CPA) lasers [3, 4] has opened a new interaction regime characterized by ultrashort and ultraintense laser pulses and steep density gradient plasmas [5]. Ultraintense laser–matter interactions are now considered for a wide range of applications including, for example, the acceleration of charged electrons and ions; the generation of novel, short-wavelength, bright radiation sources for medical applications [6]; and femtosecond crystallography [7], the production of warm-dense matter for astrophysics studies.

Indeed, most of the applications of laser–plasmas rely on the efficient production of energetic electrons driven by the interaction of ultraintense laser pulses with plasmas created from solids or gases. In fact, in these interaction conditions, laser energy is efficiently transferred to electrons generating a population of so-called fast or hot electrons. The process of fast electron generation often takes place near the critical density (the density at which the laser frequency ω_0 equals the local plasma frequency $\omega_{\rm pe}$) surface [8, 9] due to resonance absorption or vacuum heating [10], but may also arise from collision-less damping of electron plasma waves generated in the underdense plasma by laser-driven instabilities. Measurements of the angular distribution of fast electrons can be used to trace back the process and identify the originating mechanism as discussed in [11].

These hot electrons, with energies ranging from some tens of keV up to the MeV range, can penetrate the underlying cold target material, leading to inner shell ionization of the target atoms. Successive de-excitation of the ionized atoms through radiative transitions leads to the emission of characteristic K shell lines, in particular the $K\alpha$ line. As the production of energetic electrons occurs only during the laser pulse, in principle, $K\alpha$ sources with sub-picosecond pulse duration can be achieved [12]. Extensive experimental studies on $K\alpha$ sources were reported by several groups, regarding X-ray yield measurements [13–17], polarization dependent laser absorption studies [18–20], spatial [21–23] and temporal [24] characterization of the X-ray emission, and the effect of pre-pulses on the X-ray conversion efficiency [25–27].

Indeed, one of the most remarkable examples of exploitation of this efficient production of energetic electrons is the possibility of achieving inertial fusion in the so-called fast ignition (FI) approach [28]. In this case, high current, fast electron beams are required to propagate and deposit energy efficiently in near solid density collisional plasmas with consequent onset of transient magnetic fields and neutralizing plasma return current. Investigation of these mechanisms requires accurate and selective diagnostic techniques capable of probing inner regions of dense plasmas, mostly based on X-ray radiation. In view of this, a large effort is ongoing at large and smaller scale high-power laser laboratories to develop experimental techniques applicable to fast-ignition relevant experiments and planned future full scale installations like the European High Power Laser Energy Research facility (HiPER) [29].

In this chapter, I discuss experimental issues in the investigation of fast electron transport studies in solids irradiated at relativistic intensities, with particular attention to space-resolved spectroscopy relevant to both ultrashort $K\alpha$ X-ray sources and fast ignition studies. In Sect. 7.2, the basic diagnostic techniques are summarized and the limitations of these measurements are briefly described. In Sect. 7.3, an introduction is given of an alternative spectroscopic technique based upon single-photon, also known as single-hit, detection. In Sect. 7.4, an extension of this technique is presented to a novel, multipurpose X-ray imaging technique based upon a pin-hole camera equipped with a CCD detector working in single photon detection. As we will see, this technique [30] allows X-ray images to be obtained, which are resolved in energy over a large photon energy range. Finally, Sect. 7.4.2 focuses on the application of this technique to the study of X-ray fluorescence emission due to fast electron propagation in multilayer targets. Section 7.5 is dedicated to the summary of the main results and the perspectives of the field.

7.2 Basic Spectroscopy Techniques

Presently, the basic mechanisms behind this complex process of fast electron transport are mainly being investigated in laser–foil target interaction experiments. In such experiments, fast electrons generated by the laser interaction at the target interface, with energies ranging from some tens of kilo electron volt up to the mega electron volt range, can penetrate the underlying cold target material, leading to X-ray emission through fluorescence, the so-called Kα emission, via inner shell ionization of the target atoms. As a consequence, Kα emission can be considered as a signature of these electrons [32–35]. The basic process is the dependence of the fluorescence yield upon the energy of the electrons. As an example, the plot of Fig. 7.1 shows a comparison of the calculated and measured values (from [36]) of the cross section of X-ray fluorescence as a function of electron energy for Titanium.

According to this plot, the cross section has a pronounced peak between 10 and 100 keV. Above this value, the cross section decreases rapidly, reaching a minimum around 1 MeV and is basically constant for higher energies. Similar curves can be found for different materials, making it possible to estimate the expected response of each target material to a given energy of the fast electrons. Additionally, complementary information on the transport of fast electrons may be inferred from the X-ray Bremsstrahlung emission from the fast electrons undergoing collisions in the material or deflection due to the strong, self-generated magnetic fields as discussed in [37, 38].

From an experimental viewpoint, bent Bragg crystals coupled either to X-ray films or to CCD detectors are among the mostly used X-ray diagnostics

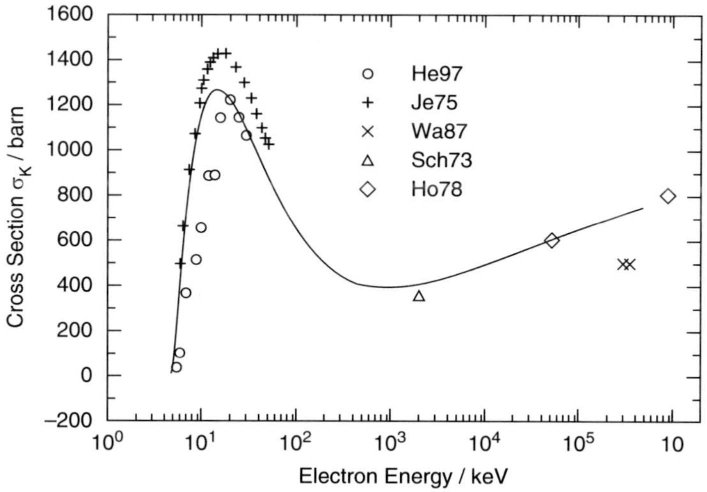

Fig. 7.1. Cross section of X-ray fluorescence as a function of electron energy showing a comparison between calculated and measured values (from [36])

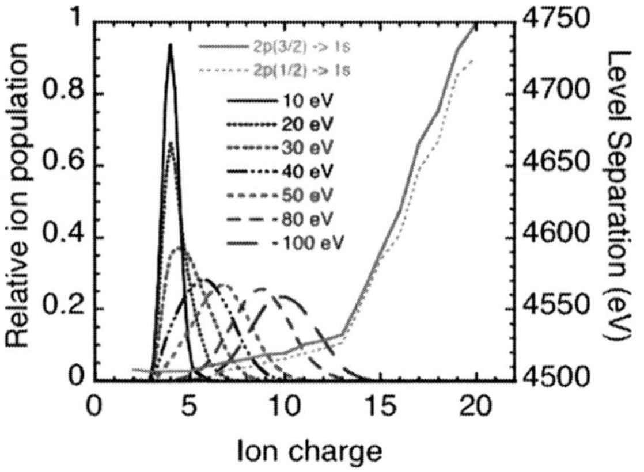

Fig. 7.2. Calculated [40] relative ion population as a function of ion charge state in solid Ti heated at temperatures ranging from 10 to 100 eV. Also shown in the same plot is the separation energy of the 2p-1s transitions as a function of the ion charge. Higher ionization stages from B-like Ti to O-like Ti are expected to emit in the range between 4,550 and 4,750 eV

for spectroscopy of laser-produced plasma X-ray emission. While allowing micrometer-scale spatial resolution to be achieved, bent crystals in the Bragg configuration suffer from some important limitations. In general, their use in relativistic laser–plasma interaction environments is made difficult by noise issues [39], especially in those configurations requiring short distances between source, crystal, and detector. Another important limitation is the small range of Bragg angles typically available. This is of a key importance when dealing with fluorescence emission from fast electrons in solids. In fact, the material from which Kα originates may undergo heating and subsequent ionization that can result in a spectral blue shift of the observed Kα line due to the reduced screening effect of the nuclear charge by the outer electrons due to ionization. This effect may result in a reduction of the collection efficiency of the spectrometer whose central wavelength is set on the unshifted fluorescence wavelength. As an example, the plot of Fig. 7.2 shows the ion population calculated as a function of the ion charge in the case of Ti for a set of electron temperatures from 10 to 100 eV. The calculation was carried out using the kinetic code FLYCHK [40].

Also shown in the same plot is the emission wavelength of the main 2p-1s transition for the different ionization stages. In the explored set of temperatures, the peak of the ion population is expected to range from Ti^{4+}(Ar-like Ti) to Ti^{10+}(Mg-like Ti). The corresponding level separation energy of the Kα emission is expected to increase from the unshifted value of 4.51 eV up to 4.53 keV. This shift is sufficient to affect the response of Bragg spectrometers as described in details in [41]. These limitations call for alternative approaches to more robust diagnostic devices, with higher flexibility in terms of the accessible spectral range and spatial resolution.

7.3 The Single Photon Detection Technique

The use of CCD detectors in the so-called single-photon regime for ultrashort and ultraintense laser–plasma X-ray spectroscopy [59, 61] is now being considered in particular in sub-PW or PW laser interaction regimes [44]. As it is well known, CCD detectors, while still limited in spatial resolution when compared to X-ray films, offer a number of advantages such as high linearity, dynamic range and quantum efficiency, as well as, with some respects, a higher ease-of-use [45, 46]. When used in the single-photon regime, a CCD detector enables the spectrum of the impinging X-ray radiation to be obtained without any additional dispersing device. This is basically due to the fact that each X-ray photon absorbed in the sensitive layer of the CCD gives rise to a signal (charge) proportional to the photon energy. Because of the large number of pixels in a conventional CCD chip, a few laser shots (ideally one) are required to get an X-ray spectrum in a large energy interval, with a typical resolution well below 10% in the energy range from 1 keV to a few tens of keV.

As anticipated above in Sect. 7.1, in principle, a CCD detector enables the X-ray spectrum of a given source to be retrieved without any external energy dispersing device, provided that an average of no more than one photon hits each pixel. In practice, because of the charge spreading around neighboring pixels, the average number of photons per pixel must be typically much less than one [47]. This condition can be achieved either by increasing the source-to-detector distance or by placing appropriate filters to attenuate the incident flux on the CCD to the required level.

Figure 7.3 shows an example of a typical raw data from a CCD detector working in a low flux regime showing individual X-ray photons. The image

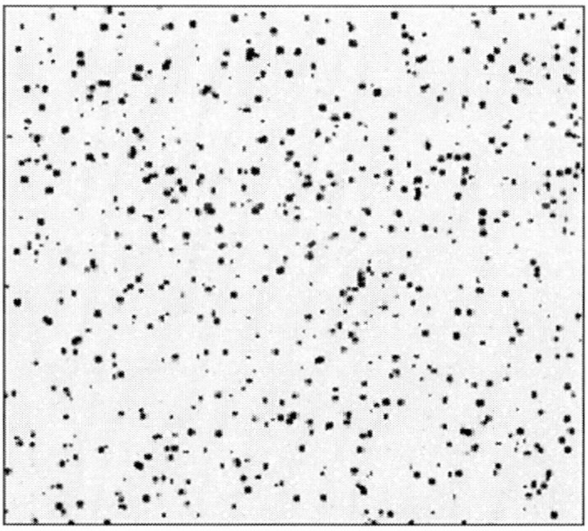

Fig. 7.3. Typical raw data showing the low flux image obtained from detection of individual X-ray photons by a CCD detector. The image shows a 512×512 pixel region of the CCD. Each X-ray photon generates a charge that is proportional to its energy

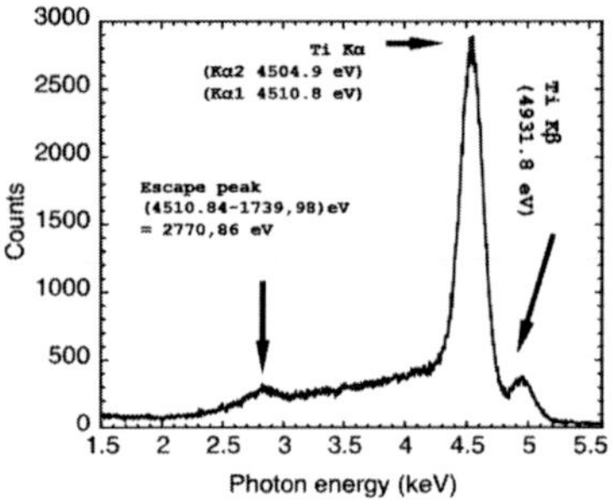

Fig. 7.4. Spectrum of X-ray emission from a 12 μm thick Ti foil irradiated at an intensity of 5×10^{17} W cm^{-2}. The spectrum was obtained from accumulation of 30 laser shots (see text for details of the irradiation conditions)

consists of a collection of spots (each with its own shape and signal amplitude) corresponding to a single photon interaction with the CCD detector. Images like this are processed and the charge corresponding to each photon event is measured and converted in incident photon energy through a calibration curve obtained using a set of known X-ray photon energies from both reference radioactive sources and well characterized laser-driven sources [48]. To obtain a full spectrum using this technique, a number of frames like the one shown in Fig. 7.3 are required to reduce the statistical noise. This is typically obtained via integration over several laser shots. More recently, because of the large area and the number of pixels available in commercial CCDs, single shot measurements are becoming possible. The plot of Fig. 7.4 shows an X-ray spectrum obtained from laser irradiation of a 12 μm thick Ti foil at an intensity of 5×10^{17} W cm^{-2} using the CNR/ILIL fs Ti:Sa laser system, which delivers up to 150 mJ in a 65 fs duration pulse. In this case, the angle of incidence was 55° and the CCD detector was looking at the target from the front side, on the horizontal plane, at 10° from the target normal axis.

Clearly visible in the spectrum are the $K\alpha$ and the $K\beta$ components of X-ray emission generated by fast electron propagation in the substrate of the Ti foil target. Also visible in the spectrum is a small feature at 2.7 keV due to a small number of events in which a $K\alpha$ photon from an excited atom of the Si substrate of the CCD array escapes from the detector sensitive area.

Recalling the image of Fig. 7.3, we observe that as this detection technique is based on a position sensitive device (typically a CCD), it also provides information on the landing position of each X-ray photon. Therefore, depending on the specific experimental setup, this information can be used to recover the angular distribution of incident X-ray photons or the spatial distribution at the source position. The following section is devoted to a description of a simple imaging device like a pin-hole camera coupled to the CCD detector to obtained full spectral and spatial information on the X-ray source. A detailed account of the instrumental aspects of the technique are given elsewhere [30, 31]. Here we give an overview of a typical experimental setup used to test this technique in a fast electron transport experiment and we make a preliminary discussion of the X-ray emission mechanisms and how they are related to fast electrons.

7.4 Energy-Resolved Imaging

As anticipated earlier, the same detection principle discussed in the previous sections can be applied to the standard pin-hole camera imaging. In fact, provided that the X-ray flux at the image plane is sufficiently low to allow single-photon identification, one can obtain a full, spectrally resolved image by collecting a large number of photons, either by accumulating images over many shots or by collecting many images simultaneously. In the following, an experiment is described in which this technique was extensively used [31] to obtain spectrally resolved images of X-ray emission from foil targets irradiated at ultrahigh intensities. The 8 TW Ti:Sa laser system at IOQ-Jena has been used to irradiate thin foil targets at relativistic intensities. The laser provided 70 fs duration pulses with 600 mJ energy at a 10 Hz repetition rate. A 45°, $f/1.2$ off-axis parabola has been used to focus the laser beam onto the target surface down to a 5 µm^2 spot, thus reaching an intensity of about 5×10^{19} W cm^{-2} (normalized vector potential $a_0 = eA_{\mathrm{L}}/m_{\mathrm{e}}c^2 \simeq 4.8$). The angle of incidence on the target was about 10°. The target was moved horizontally or vertically to ensure a fresh surface to be irradiated at each shot. The target consisted of a sandwich of layers of different Z materials and were designed to study the fast electron propagation inside the target by detecting the $K\alpha$ emission from each layer.

Two pin-hole cameras were used to image out the X-ray emission from both the front and rear side of the target. The viewing angle was about 45° with respect to the surface. Each pin-hole camera consisted of a 5 µm diameter pin-hole, bored in a 25 µm thick Pt substrate, and a back-illuminated, deep depletion CCD camera. The 1024×256 pixel CCD chip (Andor DX420) was typically cooled at −65 °C. The two pin-holes were places at about 30 mm from the target irradiation point, on both the front and rear side, and were protected from debris by using a 5 µm thick *mylar* foil. To obtain the required magnification, $M \simeq 10$, the two CCD cameras were placed outside the main vacuum chamber, in separate vacuum chambers equipped with 50 µm thick *kapton* windows (X-ray transparent). An additional Be filter was used to shield CCDs against visible light.

Cylindrical lead collimators with 10 mm thick walls and a clear aperture of about 10 mm diameter were placed between the pin-hole and the X-ray output window of the main chamber to reduce scattered photon noise inside the vacuum chamber. Also, a set of magnets (with typical magnetic field strength of about 1 T) were placed inside each tube to stop high-energy electrons. All these measures were adopted to reduce as much as possible X-rays produced by electron-impact with the tube walls and the filters. Finally, in order for the two CCD cameras to operate in the single-photon regime, up to 20 mylar foils (each one 50 μm thick) were used for each pin-hole camera to reduce the flux to the required level.

Given the pin-hole camera magnification (10×) and considering the pixel size of the two CCDs ($26 \times 26\,\mu\mathrm{m}^2$), the spatial resolution achievable by the pin-hole cameras was mainly limited by the pin-hole diameter and is therefore expected to be approximately 5 μm. More quantitative information on the performances of our imaging system in terms of spatial resolution was obtained by using Ti foils and by running the pin-hole cameras in the standard high flux condition (no energy resolution). A summary of these results is given in the following section.

7.4.1 X-Ray Imaging of Interactions With Ti Foil Targets

In Fig. 7.5, we show a comparison of front and rear side images of the X-ray source obtained at a laser intensity of $5 \times 10^{19}\,\mathrm{W\,cm^{-2}}$ for two different values of the target thickness of 5 and 25 μm. The size of the imaged area is $120 \times 120\,\mu\mathrm{m}$. The 45° observation angle was not accounted for to show full details of the uncorrected image.

We point out here that in the general case of extended, 3D emission region (e.g., propagation in thick, layered target as described below), correction of the observation angle falls in the general tomography-like reconstruction procedure and is therefore not straightforward. Even if for 5 μm target foils this would be an immediate task, and for 25 μm thick target foils this would be not straightforward. In this case, the dimension of the source is comparable with the target thickness and it is impossible to say where photons are coming from. In this case no filters were used, that is, the images are not energy resolved. The images were obtained by summing up about 100 single images. To avoid

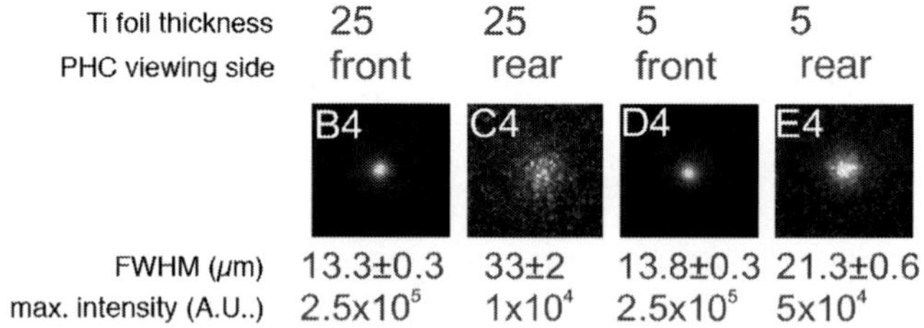

Fig. 7.5. Front and rear side images of the X-ray source obtained at a laser intensity of $5 \times 10^{19}\,\mathrm{W\,cm^{-2}}$ for two different target thickness values of 5 and 25 μm. The dimension of the images is $120 \times 120\,\mu\mathrm{m}$. Images were obtained adding up about 110 images obtained in the identical experimental conditions. In each image, the intensity is rescaled to use the entire color range. The FWHMs are also reported as measured after background subtraction from the lineouts of the source taken along vertical diameters

instrumental source broadening due to residual source displacement from shot to shot, the center of mass of a square 50×50 pixel around the expected position of the source was calculated. Possible shifts were compensated and then the images were summed up as discussed in [30].

The FWHM source size reported for each image was measured along a vertical diameter to avoid the effect of differential attenuation of the X-ray signal expected in the horizontal direction due to the angle between the observation axis and the target plane. Also shown in the image are the relative maximum X-ray intensity for the four images. Accordingly, a strong attenuation of the signal in the rear side image is found with respect to the front side emission. This attenuation can be explained as due to absorption of the X-ray radiation in the target substrate. However, as the attenuation length for Ti Kα photons in Ti is 20.3 μm, a quantitative explanation of the observed attenuation requires additional information. We also observe that the rear side images show a source size that is significantly larger than the size measured from the front side images (2.5 times in the case of the 25 μm thick target and 1.5 times in the case of the 5 μm thick target). These observations suggest a possible role of non-isotropic radiation emission mechanisms that give different contribution to the front and the rear side pin-hole cameras. Contribution of directional bremsstrahlung (see [38, 49]) to these measurements is under investigation and additional experimental evidence is given below.

7.4.2 Fast Electron Transport in Multilayer Targets

Multilayer targets are usually employed to use X-ray emission at different depths inside the target by the "tracer layers" as a marker of fast electron propagation. It is well known that layer transition in this kind of targets is expected to generate some perturbation in the fast electron propagation, due to the discontinuity, at the layer interfaces, of important parameters like material density, conductivity, etc. Nevertheless, important information on the energy deposition can still be obtained as demonstrated in recent studies (see [51] and references therein). In those studies, single energy X-ray imaging with micrometer scale spatial resolution was indeed exploited to infer substrate heating due to fast electron energy deposition.

Here we briefly show how this novel imaging technique can be used to perform multi-energy X-ray imaging with a comparable spatial resolution. To this purpose, we will show some results obtained from laser irradiation of a target consisting of three metallic layers obtained from galvanic deposition of Cr and Ni on a Fe substrate. The first layer (laser side) was 1.2 μm thick Cr, the middle layer was a 10.9 μm thick Ni, and a third layer (on the rear side) was a 10 μm thick Fe.

Energy resolving imaging was applied setting the CCD detector to work in the low flux regime. A typical raw data X-ray image shown in Fig. 7.6 consists of a small number of events (each with its own shape and signal amplitude) corresponding to a single photon interactions with the CCD detector.

Fig. 7.6. Typical raw data showing the low flux image obtained from detection of individual X-ray photons by a CCD detector coupled to a pin-hole camera. The 50 μm diameter circular area at the center of the image identifies the region where the image of the source is located. The events outside this region are due to diffuse scattering and are discarded during the analysis. Such an image is obtained from one laser shot. As in the case of Fig. 7.3, each photon generates a charge proportional to its energy. In addition, in this case the position of each event of the CCD detector is also measured to reconstruct the full image at any given photon energy

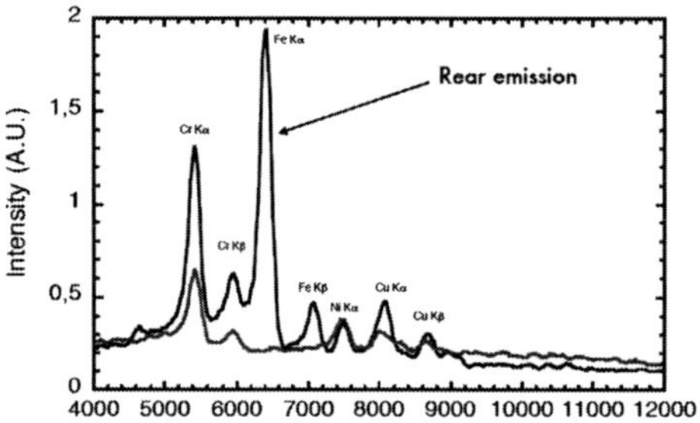

Fig. 7.7. Spectrum of the X-ray radiation detected from the front side and from the rear side of a three-layer Cr–Ni–Fe target irradiated at an intensity of $5 \times 10^{19}\,\mathrm{W\,cm^{-2}}$. The spectrum was obtained adding up all identified photon from the analysis of the low flux images obtained from 350 laser shots. The target was irradiated on the Cr side. Note that the relative intensities of the two spectra are not to scale

Each CCD acquisition (corresponding to one laser shot) is then analyzed to retrieve the total charge released and the landing position of each X-ray photon on the CCD sensitive surface. As described earlier, a dedicated reconstruction algorithm is then used to measure the charge of each photon. As a result, the total spectrum of the X-ray emission from the target can be obtained, after summing up the contribution from a few hundreds of laser shots (i.e., CCD acquisitions). Figure 7.7 shows the spectra obtained by front and rear side diagnostics (integrated over 350 shots). Table 7.1 reports the energies of the Kα and Kβ emission lines from the target materials.

From the width of the lines we find that the spectral resolution ranges from about 3.5% at 5 keV down to 2.5% at 8 keV. This value of the resolution

Table 7.1. Kα and Kβ line emission energy for the elements used in our multilayer target

Element	Kα (keV)	Kβ (keV)
Cr	5.41	5.95
Ni	7.48	8.26
Fe	6.40	7.06

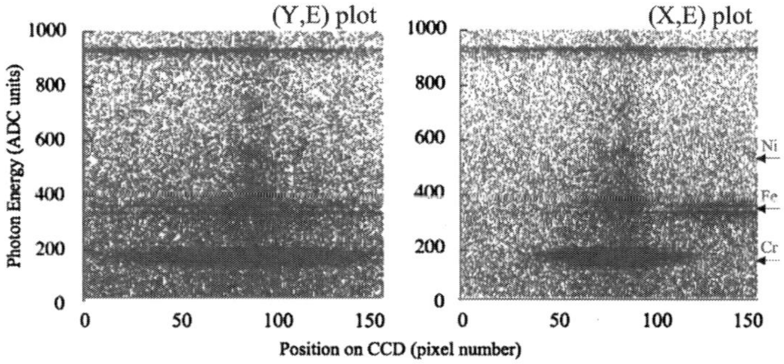

Fig. 7.8. Raw energy distribution in ADC units of all the X-ray photons plotted according to their landing position on the x–y plane of the CCD array

is mainly due to the width of the analog-to-digital converter (ADC) level distribution of the intrinsic background noise of the CCD. This is a characteristic figure of the CCD that makes the charge produced by monochromatic X-ray photons at a given energy spread over a finite width. The main Kα and Kβ line emission from the Cr layer are clearly visible on the front and rear side spectrum. The Ni Kα emission is also clearly visible from both sides of the target, while Fe Kα and Kβ are clearly visible only from the rear side spectrum. Cu emission visible in both spectra is likely to arise from impurities in the Fe and Ni substrates. We observe here that the difference in the line intensity between the front and rear side spectra are also due to the photon propagation and attenuation through the different layers. Therefore, a quantitative analysis of these relative intensities must take into account this effect as well. Each X-ray photon contributing to the spectrum of Fig. 7.7 can also be identified according to its landing position (x, y) of the CCD array as shown in the raw data plots of Fig. 7.8.

These plots show the projections on the (y, E) and (x, E) planes of the distribution of all the X-ray photons identified according to their position in the (x, y, E) space, that is, the energy resolved image of the source. Clearly visible in the plots are the contributions of the main fluorescence lines of each target layer. In most cases, the emission is centered at a common position (corresponding to the expected position of the source). In contrast, the emission around 900 ADC levels (corresponding approximately to 22.1 keV) is uniformly distributed on the (x, y) plane. In fact, this emission originates from fluorescence of materials other than the target, most likely from high Z material (Cd, Ag) present in the magnets placed between the target and the CCD to stop fast electrons from reaching the CCD. These circumstances clearly explain the powerful features of this detection techniques that enables to discriminate between signal and noise. Finally, energy resolved images of the source at any given energy can be obtained by selecting the corresponding photons from the spectral distribution, by plotting them according to their position on the CCD detector and by taking into account the geometry of the pin-hole camera imaging system, including the magnification. Figure 7.9 shows the energy resolved full imaging of the X-ray source at the Kα and Kβ line energy of the three materials, as seen by the front and rear side diagnostics. In this case, a total 1,600 single shot images were used.

The upper side of Fig. 7.9 shows the images obtained selecting X-ray photons in an energy window around a Kα line ($\pm$230 eV). In the lower side of the figure, we report the images obtained selecting X-ray photons in a spectral regions where no spectral lines are expected, that is, lower than the Cr–Kα and higher than Ni–Kα. Therefore, contribution to these images will mainly come from continuum radiation, possibly arising from bremsstrahlung emission. Higher energy photons are also present in the spectrum; however, for these energetic photons, the 2.5 μm thick Pt pin-hole substrate is partially transparent and the contrast of the pin-hole image is consequently reduced.

In general, a careful analysis of the spectrally resolved images of Fig. 7.9 can be used to identify all contributions from each layer of the target and to reconstruct the propagation of the fast electron beam through the whole target. In fact, by measuring the exact position and size of each emission component, one can reconstruct all possible sources of emission. Thanks to this novel imaging technique, the photon energy and spatial data as briefly described above provide sufficient information to enable an evaluation of the contribution of each layer of the target to the X-ray emission.

Additional information concerning the fast electron distribution can be obtained in such experiments by direct measurement of forward escaping electrons using a calibrated stack of radiochromic films [52] that can provide information of the angular and energy distribution of fast electrons. These measurements were performed in the same experiment described above [31] and the summary of those measurements is reported in Fig. 7.10 and reveal a distribution that is consistent with a relativistic Maxwellian distribution with a characteristic temperature of 160 keV.

Such an experimental characterization is a necessary step to carry out a detailed comparison of emission properties as measured experimentally with the corresponding quantities as calculated by numerical models capable of describing transport and energy deposition of fast electrons in matter and consequent emission of characteristic X-ray emission. A possible modeling approach of fast electron transport experiments is given here, where the above results on Kα imaging were interpreted using the hybrid code PETRA [53] to

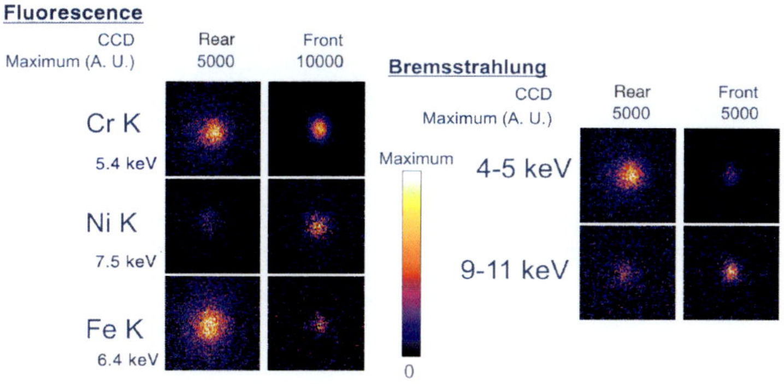

Fig. 7.9. (*Left*) Energy resolved images of the Kα sources of Cr, Ni, and Fe as seen from the rear and the front of the target obtained selecting all the photons within a spectral window of $\pm 230\,\mathrm{eV}$ around the central Kα energy. (*Right*) images of the source obtained selecting the photons in two regions where most contribution is expected to come from bremsstrahlung emission. The images are not corrected for the absorption of the Kα radiation through target layers, filters, etc. The maximum intensity value of the color scale is reported in each case (5,000 or 10,000). The dimension of each image is $180 \times 180\sqrt{2}\,\mu\mathrm{m}^2$

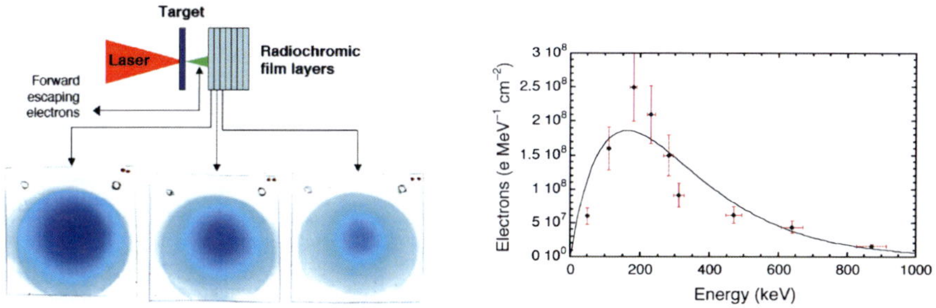

Fig. 7.10. *Left*: Scan of the first three radiochromic film layers placed behind the target, along the laser propagation direction showing the raw cross-section image of the forward escaping fast electrons generated in the same condition as the ones of Fig. 7.9. *Right*: Spectrum of the electrons retrieved using all radiochromic film layers according to [52]. Also shown in the plot for comparison is the curve corresponding to the relativistic Maxwellian distribution at 160 keV

calculate expected X-ray fluorescence from the target assuming a given fast electron distribution. PETRA is a 3D hybrid code for transport of relativistic electrons in dense plasma: fast electrons are coupled to low temperature electron background fluid by collective fields (PIC procedure) and collisions (Monte Carlo procedure). In the code, fast electrons are injected in the target as a beam with a Maxwellian relativistic distribution and local temperature taken from [54] at different regions of the focal spot. The distribution in space and time for the injected fast electron beam is Gaussian. The temperature is obtained from the measurements of Fig. 7.10 (right) after Bethe–Bloch

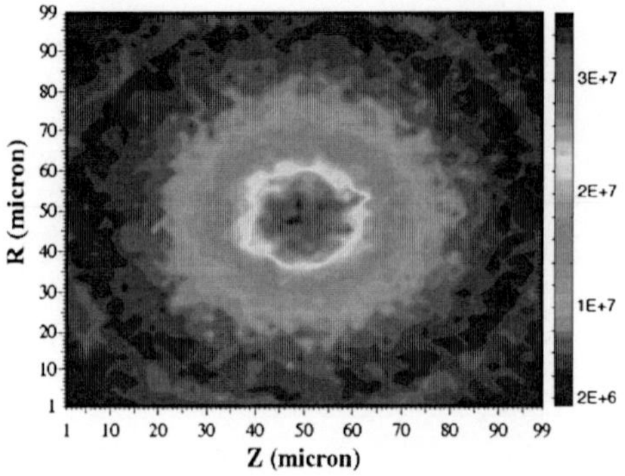

Fig. 7.11. Fast electron beam density after propagation in a 30 μm thick Fe layer calculated using the hybrid code PETRA (from [31]). The initial fast electron beam diameter at the target surface is 2.6 μm and the pulse duration is 100 fs

calculation to retrieve the electron distribution before propagation in the target. The pulse duration is 100 fs and the diameter of the fast electron beam at the target surface is assumed to be 2.6 μm to match the focal spot size of our experimental set-up. The target substrate is described by its specific heat as an interpolation of the Fermi–Dirac electronic specific heat and the ideal gas specific heat and by its resistivity based on a modified Spitzer model [55]. Finally, Kα emission is calculated from the impact ionization cross-section.

A simulation was carried out to model the fast electron beam density after propagation in a 30 μm thick Fe layer and the result is displayed by plot of Fig. 7.11.

According to this plot, the fast electron beam undergoes a significant diffusion in the target that makes the beam size increase by a factor of two or more at the exit of the target. In fact, Kα emission calculated by the code during this fast electron propagation has a source size with a main component of approximately 10 μm FWHM, while lower intensity tails of the emission extend up to a 50 μm from the axis. These calculation are now under extensive test and quantitative comparison with the measurements described above. Recent additional modelling work also suggest that strong self-induced magnetic fields may occur at interfaces in layered targets which may lead to additional collimation. The difference between the calculated size of the fast electron beam of Fig. 7.11 and the size inferred by the images of Fig. 7.9 may therefore be accounted for by such collimating magnetic fields.

7.5 Summary and Conclusions

In summary, the main issues were presented here concerning experimental investigation of fast-electron transport in solids irradiated at ultra-high intensity using high-power, femtosecond laser pulses. A discussion was given on the limitations of basic spectroscopic diagnostic techniques based upon well established Bragg crystal spectrometers. Then an account was given of the results of a novel experimental technique for the full imaging with spectral resolution of the X-ray emission from laser irradiated targets. The technique, based upon a pin-hole camera equipped with a CCD detector operating in the so-called single-photon regime, allows the image in any given spectral window in the keV range to be obtained, provided the X-ray response of the CCD has been previously characterized. Furthermore, the absolute photon flux in each selected energy range can be retrieved once the collection efficiency of the setup and the detector QE are considered. The use of this technique was reported for the study of the propagation of fast electrons inside multilayer targets, whose fluorescence emission from each layer due to fast electron propagation can be imaged with a spatial resolution of about 5 μm and an average spectral resolution of 3% in the spectral range below 20 keV. The technique "as it is" has currently a "multi-shot" basis due to the need for a statistically significant number of photons to be detected. The number of shots could be reduced by using a higher magnification, which would enable to collect a greater number of photons per shot, while satisfying the single-photon condition on the CCD detector. Another possibility is to exploit large area CCDs or similar position sensitive devices in combination with a pin-hole array to collect more than one "single-photon image" on the same CCD detector.

Acknowledgements

We acknowledge financial support by the MIUR SPARX and MIUR-FSRIS project "Impianti Innovativi multiscopo per la produzione di radiazione X" and by the INFN project PLASMON-X. Access to the IOQ installation was supported by LASERLAB. We thank F. Cornolti from the Physics Department of the University of Pisa for fruitful discussion and suggestions on the preliminary modeling of the data.

References

1. L. A. Gizzi, et al., Phys. Rev. E **49**, 5628 (1994)
2. D.Giulietti, L. A.Gizzi, Rivista del Nuovo Cimento **21**, 1 (1998)
3. M. Perry, G. Morou, Science **264**, 917 (1994)
4. D. Stickland, G. Mourou, Opt. Commun. **56**, 219 (1985)
5. M.M. Murnane, et al., Phys. Rev. Lett. **62**, 155 (1989)
6. J.C. Kieffer, et al., App. Phys. B **74**, S75 (2002)
7. A. Rousse, et al., Rev. Mod. Phys. **73**, 17–31 (2001)
8. S.C. Wilks, W.L. Kruer, IEEE J. Quant. Elect. **33**, 1954 (1997)
9. P. Gibbon, E. Frster, Plasma Phys. Control. Fusion **38**, 769 (1996)
10. F. Brunel, Phys. Rev. Lett. **59**, 52 (1987)
11. L. Labate, et al., App. Phys. **B 86**, 229 (2006)
12. J. Limpouch et al., Laser and Part Beams **22**, 147 (2004)

13. J. Yu et al., Phys. Plasma **6**, 1318 (1999)
14. L.M. Chen, et al., Phys. Plasma **11**, 4439 (2004)
15. F. Ewald, et al., Europhys. Lett. **60**, 710 (2002)
16. C. Reich, et al., 888E. Frster, Phys. Rev. Lett. **84**, 4846 (2000)
17. M. Hagedorn, et al., App. Phys. **B 77**, 49 (2003)
18. U. Teubner, et al., Phys. Rev. Lett. **70**, 794 (1993)
19. B. Soom, et al., J. Appl. Phys. **74**, 5372 (1993)
20. U. Teubner, et al., Phys. Rev. E **54**, 4167 (1996)
21. G. Pretzler, et al., App. Phys. Lett. **82**, 3623 (2003)
22. D.C. Eder, et al., App. Phys. *B 70*, 211 (2000)
23. C. Reich, et al., Phys. Rev. E **68**, 056408 (2003)
24. T. Feurer, et al., Phys. Rev. E **65**, 016412 (2001)
25. A. Rousse, et al., Phys. Rev. E **50**, 2200 (1994)
26. T. Schlegel, et al., Phys. Rev. E **60**, 2209 (1999)
27. C. Ziener, et al., Phys. Rev. E **65**, 066411 (2002)
28. M. Tabak, et al. Phys. Plasma **1**, 1626–1634 (1994)
29. HiPER Technical Design Report, http://www.hiper-laser.org
30. L. Labate, et al., Rev. Sci. Instrum. **78**, 103506 (2007)
31. L. A. Gizzi, et al., Plasma Phys. Contr. Fusion, **49**, B211 (2007)
32. H. Nishimura, et al., J. Quantum Spectr. Radiat. Transf. **81**, 327 (2003)
33. R. Freeman, et al., J. Quant. Spectr. Radiat. Transf. **81**, 183 (2003)
34. D. Batani, et al., Phys. Rev. E **65**, 066409 (2002)
35. K. Yasuike, et al., Rev. Sci. Instrum. **72**, 1236–1240 (2001)
36. F. Ewald, et al., Europhys. Lett. **60** 710–716 (2002)
37. P.A. Norreys, et al., Phys. Plasma **6**, 2150 (1999)
38. Y. Sentoku, et al., Phys. Plasma **5**, 4366–4372 (1998)
39. C. Stoeckl, et al., Rev. Sci. Instrum. **75**, 3705–3707 (2004)
40. H.-K. Chung, et al., High Energ. Phys. Nuc. **1**, 3 (2005); Skobelev, et al., Rev. Sci. Instrum. **69**, 4049–4053 (1998)
41. J.A. King, et al., Rev. Sci. Instrum. **76**, 076102 (2005)
59. L.A. Gizzi, et al., Phys. Rev. Lett. **76**, 2278–2281 (1996)
61. L. Labate, et al., Appl. Phys. B **86**, 229–233 (2007)
44. H.S. Park, et al., Rev. Sci. Instrum. **75**, 4048–4050 (2004)
45. S.M. Gruner, et al., Rev. Sci. Instrum. **73**, 2815–2842 (2002)
46. J. Howe, et al., Rev. Sci. Instrum. **77**, 036105 (2006)
47. G.W. Fraser, X-ray detectors in astronmy, Cambridge University Press (Cambridge, 1989)
48. L. Labate, et al., Nucl. Instrum. Meth. A, **495**, 148–153 (2002)
49. Z. M. Sheng, et al., Phys. Rev. Lett. **85**, 5340 (2000)
50. F. Zamponi, et al., Rev. Sci. Instrum. **76**, 116101 (2005)
51. D. Batani, Laser Part. Beams, **20**, 321–336 (2002)
52. M. Galimberti, et al. Rev. Sci. Instrum. **76**, 053303 (2005)
53. X. Honrubia, et al., Laser Part. Beams **22**, 129–135 (2004)
54. F. Beg, et al., Phys. Plasma **4** (1997)
55. A. Benuzzi, et al., Phys. Plasma **5** 2410 (1998)

8

High Field Photonics in Laser Plasmas: Propagation Studies, Electron Acceleration, and Nuclear Activation With Ultrashort Intense Laser Pulses

Antonio Giulietti and Andrea Gamucci

Summary. Ultrashort intense laser pulses, when focused, allow to obtain photon densities as high as 10^{27} phot cm^{-3} and field amplitudes exceeding 10^{12} V cm^{-1}. This opens the exciting scenario of high field photonics in laser-produced plasmas. Millimeter size laser driven accelerators have been set up, which deliver efficiently relativistic electron bunches with a table top equipment. Such electron bunches are already suitable to be used for a variety of applications, including cancer therapy. Alternatively, they can support, via bremsstrahlung in a dumper, an intense source of gamma-rays whose spectral component in the Giant Dipole Resonance of nuclei allows photonuclear activation of radio-isotopes. To further progress in this field, a detailed knowledge and a systematic control of the laser pulse propagation is required. The paper shortly introduces and reviews relevant achievements in this domain.

8.1 Introduction

Chirped pulse amplification (CPA) technology [1] allowed laser systems to achieve unprecedented e.m. field intensities with focused femtosecond pulses. Powerful femtosecond lasers, coupled with suitable focusing optics, can shoot into matter optical "bullets" whose transverse size (the spot size) is comparable with the longitudinal size (the pulse length) and whose photon density is of the order of 10^{27} phot cm^{-3}. The oscillating electric field in the bullet volume can exceed 10^{12} V cm^{-1}, much higher than atomic fields. It corresponds to laser intensities $>10^{21}$ W cm^{-2}. When interacting with matter, such pulses are able to ionize atoms in a time of the order of a single optical cycle. Free electrons are then moved by the oscillating electric field to relativistic quiver velocities. This is the new exciting scenario of high field photonics that can be investigated in dense plasmas produced by the laser itself [2,3]. The interaction is obviously highly nonlinear and a wide class of new phenomena can be studied, starting from the pulse propagation.

One of the most striking results in this investigation has been the evidence of propagation of intense femtosecond pulses in plasmas whose density is well above the critical density (i.e., the density above which propagation is not allowed by the classical linear theory) [4]. Though the interpretation of this effect is still controversial, it can have important consequences and applications. On the other hand, the propagation of intense femtosecond pulses in underdense plasmas has to be carefully investigated in view of relevant applications. Besides the nonlinear aspects, a critical point is the presence of precursors of the femtoseconds pulse, which can modify the medium before the arrival of the main pulse and modify consequently its propagation. These aspects are particularly sensible for the use of intense femtosecond laser pulses to accelerate electrons in plasmas.

When focused in a suitable underdense plasma, the e.m. fields of such pulses are able to support electron waves of very large amplitude, whose longitudinal electric field can accelerate free electrons up to relativistic electron energies with local acceleration fields several orders of magnitude higher than in ordinary accelerators [5]. A number of schemes for laser-driven acceleration of electrons in plasmas have been proposed and studied, some of which successfully tested. New experimental records have been reported in the recent literature about the maximum electron energy achieved, the minimum energy spread, as well as maximum collimation, stability, and so on [6–8]. These records are in general obtained with lasers of outstanding performances and/or with very sophisticated methods hardly applicable in practical machines. A special effort is devoted to preform conditions in which the focused laser pulse can propagate at the peak intensity beyond the length (Rayleigh length) allowed by the diffraction of the focusing optics. Partially successful techniques include capillary discharges and preformed plasma channels.

The scientific and technological path towards the realization of competitive laser-driven accelerators still demands a considerable up-grade of many performances, including energy spread, collimation and pointing stability of the electron bunches, high repetition rates with reproducibility of the bunch parameters. Though we will briefly review the main results obtained in this path so far, we will focus on the possibility of having *now* tabletop sources of relativistic electrons driven by ultrashort, intense laser pulses. In particular, there is a specific interest to sources of electrons suitable for an efficient activation of radionuclides and radiotherapy of tumors. To this purpose, extreme performances in terms of energy, energy spread, collimation, and pointing stability are not required. On the other hand, electrons of energy of tens of MeVs, fairly collimated, with high average current (i.e., high charge per bunch at a given repetition rate) have to be produced efficiently and reproducibly. Recent experiments proved that such requirements are close to be achieved with tabletop lasers (pulse peak power $\leq$10 TW) and gas-jets. Detailed studies on the laser pulse propagation in a gas (plasma) on the acceleration mechanism and on the radio-activation method gave a basic contribution in obtaining such promising results. We will report and discuss those studies and their main results to address further investigations and technological improvements towards the realization of a source for practical uses.

Although this work is restricted to electron acceleration, it has to be mentioned here for completeness that an increasing amount of work is currently devoted (often in the same laboratories) to the study of proton and light ion acceleration with laser pulses, also in view of future medical applications. Among a number of recent works on this subject, see for example [9] and [10].

8.2 Studies on Laser Pulse Propagation

Propagation in a medium of a coherent optical wave packet whose longitudinal and transverse sizes are both of a few wavelength and whose field amplitude can induce relativistic motion of electrons is a novel challenging topic to be investigated in the general field of the so-called *relativistic optics* [11]. Theory and simulation have been applied to this problem for a few decades. A number of experiments have been performed since ultrashort intense laser pulses became available in many laboratories.

Relativistic transparency of overdense (density above critical density) plasmas is likely to be the most striking effect as long as propagation of laser light of pulsation ω_L is not allowed, by linear nonrelativistic theory, in a plasma whose density n_e exceeds the critical density $n_c = (m_e/4\pi e^2)\omega_L^2$. Overdense propagation has been clearly observed, however, in the interaction of ultrashort intense laser pulses with thin foils. The interpretation of the experimental data is still controversial. Propagation in underdense plasmas may appear not so exciting; nevertheless, its study reveals unexpected features, essential for any kind of applications, including particle acceleration.

In general, it has to be noticed that data from experiments have to be compared with data from numerical simulation with great care and prudence. As a matter of fact, most of the numerical codes describe propagation of the ultrashort laser pulse (1) without any pre-pulse and (2) in a medium already ionized. These two conditions are usually far from being fulfilled in a real experiment. In the following, we will show how precursors of the main laser pulse and ionization of the medium can deeply change the propagation of an ultrashort intense laser pulse.

8.2.1 Propagation in Overdense Plasmas

In the case of nonrelativistic laser intensity, linear theory does not allow propagation in overdense plasmas, namely when $\omega_L < \omega_{pe} = e(4\pi/m_e)^{1/2}n_e^{1/2}$. In the extreme case of ultra-relativistic laser intensity ($a_0 \gg 1$), the cutoff frequency for propagation drops from ω_{pe} down to $\omega_{pe}/(1+a_0^2)^{1/4}$ [11], where $a_0 = eA/m_e c$ is the "dimensionless amplitude" of the laser field. Then, in order for the propagation to occur at plasma density appreciably higher than the ordinary critical density, $a_0 \gg 1$ is needed. This is also the case of overdense thin plasma layers (as proved by simulation [12]) whose thickness exceeds the skin penetration depth of the e.m. wave. Theoretical background and basic

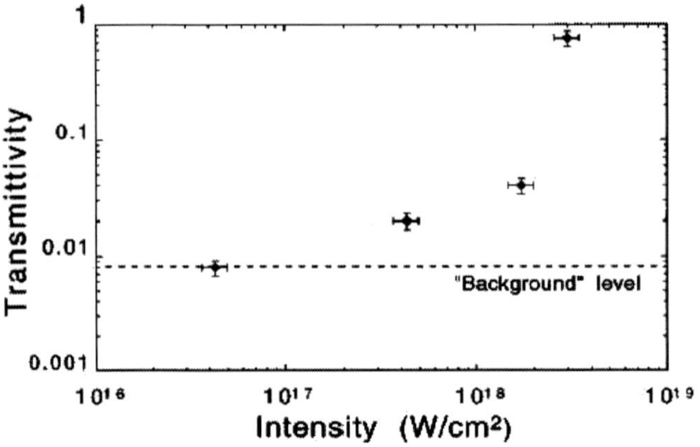

Fig. 8.1. Overdense plasma transmittivity vs. laser intensity on target for a focused Ti:Sa laser pulse (30 fs, 800 nm) [4]

numerical concepts on overdense plasma propagation based on Helmholtz solver can be found in [13].

Surprisingly, transparency to the laser light was observed in a 0.5 μm thick plasma slab (much thicker than the skin depth) obtained by ultra-fast ionization of a plastic foil at a density orders of magnitude higher than n_c [4]. The effect was observed at laser intensities corresponding to $a_0 \approx 1$. Figure 8.1 shows the measurements of plasma transmittivity in that experiment vs. the intensity on target of a focused Ti:Sa laser pulse (30 fs, 800 nm).

This striking effect was found to be definitely dependent on the pulse over pre-pulse power ratio, the so-called *contrast ratio* (see also next paragraph). Only when the pre-pulse intensity on target was low enough to avoid pre-ionization of the foil, the effect could be observed. If the foil was ionized by the pre-pulse, prior to the arrival of the main pulse, this latter could not propagate. This observation suggested a model explanation based on the onset of very high magnetic field due to ultra-fast ionization [14]. Other tentative explanations include ponderomotive modification of the density profile [15] and transverse currents driving resonant transport [16, 17].

Further relevant features of that phenomenon were observed [4], including (1) spectral shift and broadening of the transmitted laser light clearly proving the occurrence of ultra-fast ionization of the foil, (2) spatial filtering of the transmitted laser pulse due to the different transmittivity across the pulse section depending on its local intensity, according to the measurements shown in Fig. 8.1. Similar to the observed spatial filtering, a reshaping of the temporal profile of the laser pulse could also be expected. Though the spectral broadening is consistent with a pulse shortening, data were not enough for a complete reconstruction of the pulse time-shape after propagation. We recall that strong pulse reshaping is expected after ultra-relativistic laser–foil interaction [18, 19].

Soon after, overdense propagation was also observed with longer laser pulses (of the order of 500 fs) and intensities up to $1.5 \times 10^{19}\,\mathrm{W\,cm^{-2}}$ [20].

8.2.2 Propagation in Underdense Plasmas and Pre-pulse Action

Any kind of basic studies and practical applications involving ultrashort intense laser pulses requires a preliminary knowledge of the regime of propagation of such pulses in the given medium. Nonlinear relativistic theory [11] can provide the correct description for the propagation of very high e.m. fields. Numerical (most frequently "particle in cell" or PIC [21]) simulation can in turn apply those concepts to complex conditions, including nonuniform field and inhomogeneous plasma density distributions. However, as we have indicated before, the interaction with the medium of the laser emission prior to the main pulse (the "precursors" or "pre-pulse") and the ionization of the medium are usually not included in the numerical codes, which describe directly the interaction of the ultrashort intense pulse with a fully ionized plasma. This makes experimental results often surprising with respect to the numerical description of the process, as already shown in the case of overdense laser–matter interaction (see Sect. 8.2.1). First, even in the cases in which the pre-pulse action can be ignored, the ionization of the medium by the main pulse can deeply change its propagation due to self-modulation and self-defocusing caused by the fast change in the refractive index or generation of intense local electric and magnetic fields [14,22]. Second, very frequently, the "initial status" of the medium with respect to the intense pulse propagation has been "prepared" by the early interaction with the pre-pulse.

Numerical codes based on multiphoton or tunnel ionization by the laser pulse [23–25] can provide the map of the dilute medium at its different degrees of ionization during and after the laser pulse propagation. As for the pre-pulse action, the basic parameter to be considered is the "power contrast ratio", that is, the ratio between the peak power of the ultrashort pulse and the power of the pre-pulse. However, this single parameter cannot describe the complex action of the "precursor" laser light on the medium. It is useful to consider three different kinds of pre-pulse. The first one is a sort of early "replica" of the main pulse due to the imperfect cut of the previous pulses produced by the laser oscillator, typically running at repetition rate $\leq$100 MHz. It makes the closest precursor of this kind to precede the high power pulse at least 10 ns; further, ordinary electro-optical cutting of this precursor reduces its energy to an extremely low level, so it can be usually ignored. The second kind of pre-pulse is due to the "amplified spontaneous emission" (ASE) generated in the amplifier chain. This one has a typical duration of a few nanosecond and can carry an amount of energy comparable with the main pulse, though at much lower power rate. Third, a "picosecond pedestal", which always precedes the main pulse due to the imperfect compression of the stretched amplified pulse in the CPA system. This kind of precursor has to be monitored and its effect on the medium, just before the arrival of the main pulse peak, carefully tested.

Recently, a systematic experimental study has clearly proved the effect on a gaseous medium of both ASE and picosecond pedestal prior to the arrival of the ultrashort intense laser pulse [26]. The study has been based on sequences of electron density maps obtained from optical interferograms with femtosecond resolution and has been supported by numerical simulation of the ionization of the medium.

The experiment was performed at the SLIC laser facility in the Saclay Centre of CEA (France) with the 10 TW UHI10 laser system delivering 60 fs, linearly polarized, laser pulses at 800 nm from a Ti:Sa system operating in the CPA mode. The precursors of the main pulse were carefully monitored by a third order cross-correlator [27], supplying both the nanosecond and picosecond scale contrast of the pulse. The ASE was found to last about 1 ns before the main pulse with a contrast ratio of $\approx 10^6$, while the picosecond pedestal had a contrast ratio between 10^4 and 10^3. The propagation was studied by focusing with an $f/5$ off-axis parabola, in a Helium gas-jet (500 μm from the nozzle, density of 10^{19} at cm^{-3}), the laser pulse at a nominal intensity of 3×10^{18} W cm^{-2}, corresponding to a moderately relativistic field of normalized parameter $a_0 = 1.2$. The focus of the ASE was found not to lie in the same position as the focus of the CPA pulse; the latter being located about 400 μm toward the laser. The laser spot (13 μm FWHM, $M^2 = 3.3$) was imaged from the transmitted light channel, in vacuum, with an $f/5$ lens on a CCD camera to get its intensity distribution. A fraction of the femtosecond pulse was doubled in frequency and used as an optical probe perpendicular to both the main pulse propagation axis and the gas flow axis. The probe pulse was timed with respect to the main CPA pulse by means of an optical delay. High quality interferograms were obtained in a Mach–Zehnder configuration [28]. Reproducible series of interferograms taken at different delays showed the evolution of the gas ionization within steps of few hundreds of femtoseconds during the laser pulse propagation.

The Mach–Zehnder configuration is flexible in terms of magnification and field of view but requires the beam to be separated in two pathways far from the region of interest: one split beam passes through the plasma while the other propagates undisturbed in vacuum, and then the beams recombine. This configuration needs accurate alignment but has the advantage that the two arms can be widely separated spatially. Mach–Zehnder interferometer can also be used to obtain shadowgraphic images of the plasma, provided that the unperturbed beam is blocked [28]. The output of this detection technique is a two-dimensional fringe pattern of the plasma at the probing time. Changing the delay between the laser pulse entrance in the medium and the probing time, it is possible to follow the plasma evolution starting from its first creation. Once the interferogram is recorded, numerical de-convolution of the local fringe shift into the phase difference produced on the probe front by its propagation in the perturbed medium is needed. This work has been done

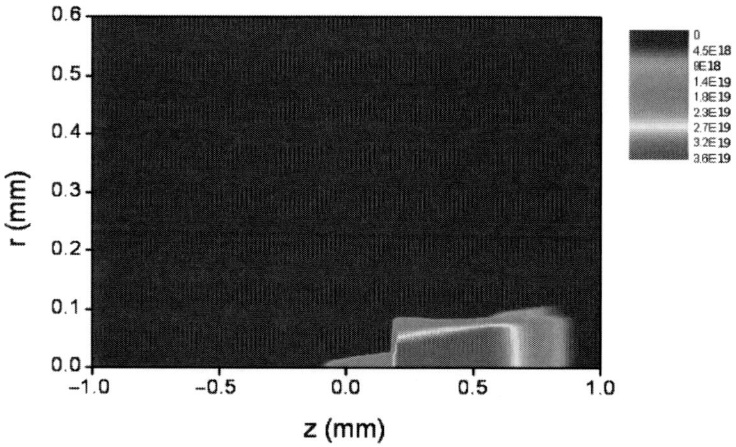

Fig. 8.2. The electron density map obtained by numerical simulation of an ultrashort intense pulse preceded by a picosecond pedestal that propagates in helium. The snap-shot refers to 1.5 ps after entrance in the gas. Laser comes from RHS (see also [26])

for a long time with Fast Fourier Transform codes. Recently a new method based on Continuous Wavelet Transforms achieved higher accuracy and spatial resolution, mostly in the case of small density perturbations [29]. From the phase difference map Abel inversion provides the electron density map $n_e(x, y)$ under the assumption of axial symmetry of the density distribution [30, 31]. A few years ago, a new algorithm that exploits the expansion of $n_e(x, y)$ in a truncated Legendre polynomial in the azimuthal angle has been successfully proposed in order to generalize the Abel inversion even to electron density distributions that slightly differ from axial symmetry [32, 33].

Figure 8.2 shows an example of the electron density map obtained by numerical simulation, in this case 1.5 ps after the entrance of the ultrashort pulse in the gas. The laser is coming from the right hand side. The picosecond precursor was included in the code, while the ASE was not.

The experimental data obtained at the same time (Fig. 8.3) show similar electron density distribution, except for the preformed plasma due to the local early gas breakdown induced by the nanosecond ASE pre-pulse (visible in the left part of the ionized region). The investigation was completed with imaging and spectroscopy of the transmitted laser pulse after the propagation in the gas.

The study confirmed that the action of the precursors of the main pulse has to be carefully considered in the ultrashort intense laser pulse interactions. Nevertheless, a regime of rather stable propagation [34] of a laser pulse of tens of femtoseconds was found in a broad window of the laser intensity/medium density diagram, in which (1) the ionization occurs in one or a few optical cycles, avoiding effects of self-phase modulation and defocusing for most of the

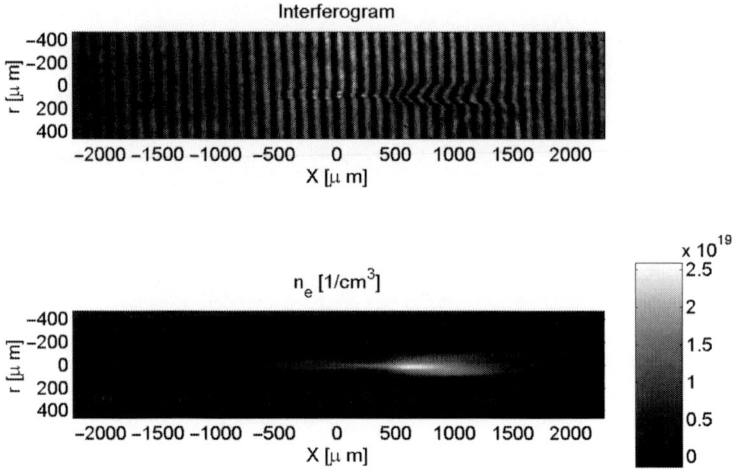

Fig. 8.3. Experimentally observed electron density map (*bottom*) obtained from the interferogram (*top*). Conditions are the same as for the simulation snap-shot of Fig. 8.2 (see also [26])

pulse; (2) intensity remains at a level $a \approx 1$, which avoids strong relativistic effects.

It has to be pointed out that ultra-high intensity effects, including relativistic self-focusing and pulse shaping, can help in some particular cases, like electron acceleration driven by laser pulses in plasmas (see also next Sect. 10.3). For this kind of purpose, also the precursors of the main laser pulse can be profitably used. In Sect. 8.2.3, we will show how nanosecond laser pulses comparable to the ASE pre-pulse can produce plasma channels able to guide the main pulse for accelerating paths much longer than the Rayleigh length. An outstanding examples of the use of pre-pulse to improve the electron acceleration process has been shown recently by applying a magnetic field of 0.2 T along the accelerating path of a 12 TW, 40 fs laser pulse [35]. The combined action of ASE and magnetic field allowed to considerably increase charge, collimation, and shot-to-shot reproducibility of the accelerated electron bunches.

8.2.3 Preformed Pulse-Guiding Channels in Plasmas

The propagation of a laser pulse focused in a plasma at high intensity is affected by several linear and nonlinear phenomena, which can modify its amplitude and therefore its possibility to travel in the medium maintaining its original interaction conditions. The main limitation arises from optical diffraction that enables the propagation at the maximum intensity only over the Rayleigh length Z_{R}:

$$Z_{\mathrm{R}} = \pi w_0^2/\lambda, \tag{8.1}$$

where λ is the laser wavelength and w_0 is the "waist" size. This latter depends in turn on the parameter $F/n = f/w_i$ of the focusing optics: $w_0 \approx f\lambda/\pi w_i$ for a beam with diameter w_i focused by an optics with focal length f. It follows that for a tightly focused laser pulse, Z_{R} is limited to small values ($Z_{\mathrm{R}} \approx 100\,\mu\mathrm{m}$ for a diffraction limited Ti:Sa laser beam focused in a $5\,\mu\mathrm{m} - 1/e$ radius – spot). This drawback affects, for example, the majority of the experiments aimed to obtain efficient electron acceleration in plasmas via laser wakefield (LWF) mechanisms (which will be introduced in the next section), in which the high-energy gain is directly dependent on the possibility for the laser pulse to excite high-amplitude plasma waves in its wake via its ponderomotive force. The loss of intensity due to diffraction over Z_{R} cannot be overcome by focusing the electromagnetic pulse in wider waists, because it would imply to use laser systems more powerful and thus less affordable to get the same intensity on target. Though nonlinear phenomena like self-focusing and self-guiding of the pulse can, in particular conditions, allow for intense propagation over paths much longer than Z_{R} (see Sect. 10.3), there is a need for achieving such a longer intense propagation also in more general conditions. One way to solve this propagation issue consists in shaping the plasma medium traveled through by the short laser pulse as a hollow-channeled wave guide that can counterbalance optical diffraction [36]. Plasma channel guiding was first demonstrated by focusing with an axicon a first laser beam on a gaseous high-Z target and exploiting the hydrodynamic channel density depletion to guide a second beam over more than 20 Z_{R} [37]. It is worth noting that the use of low-Z gases has enhanced the possible applications of such guiding technique, as a fully ionized medium can provide a more efficient propagation of an intense laser pulse in the channel avoiding energy losses in further ionization processes. The plasma channel is shaped in a way that the plasma index of refraction $\eta = (1 - \omega_{\mathrm{p}}^2/\omega^2)^{1/2}$, in which ω_{p} is the plasma frequency and ω is the laser frequency, has a local maximum on the propagation axis, and decreases radially towards the edges. In this way, the overall effect on a traveling intense pulse is the same of a continuously converging optical medium, which refocuses the pulse as it optically diffracts. In terms of plasma electron density n_{e}, it implies that its spatial distribution has a minimum on the longitudinal axis, with a value lower by Δn_{e} with respect to the outer walls. It turns out that in these conditions a laser pulse can be guided provided the following relation between w_0 and Δn_{e} holds:

$$w_0 = \left(\frac{r_{\mathrm{ch}}^2}{\pi r_{\mathrm{e}} \Delta n_{\mathrm{e}}}\right)^{\frac{1}{4}}, \tag{8.2}$$

where r_{ch} is the plasma channel radius and r_{e} is the classical electron radius [38].

Optical guiding in preformed plasmas has been extensively investigated in experiments mainly oriented to demonstrate the production of relativistic electrons in LWF-related schemes. Plasma channel formation has been pursued with a variety of means, ranging from the use of hydrodynamic and shock-wave driven density depletion in gas-jets to the employment of gas-filled capillaries and the use of self-guiding of relativistically intense laser pulses. Production of plasma channels in gas-jets has been widely investigated in the tuning of the interplay between laser pulse and target medium. In this framework, an "ignitor–heater" technique was first developed at Lawrence Berkeley National Laboratory (USA), which makes use of a first short pulse (<100 fs) to ionize a millimeter nitrogen or hydrogen gas-jet and of a second longer pulse (160 ps) to heat the existing plume via inverse bremsstrahlung [39]. This technique was introduced also to show how a double pulse process can save the energy in generating plasma channels with respect to a single pulse technique.

An alternative way to preform a channeled plasma consists in exploiting the nanosecond precursor that usually precedes a short femtosecond pulse in the output of a multi-terawatt laser system. In fact, the amplified spontaneous emission (ASE) pedestal has typically an intensity 10^6–10^{10} times lower than the main pulse, which, however, can be sufficient to ionize a gas-jet or a solid target. This drawback can be turned into a benefit assuming that this long precursor can prepare the plasma channel for the short pulse propagation.

An experiment was successfully performed to study the plasma shaping properties of a nanosecond, "ASE-like", pulse: reproducible hollow-channels were produced in a helium gas-jet at intensities around $10^{14}\,\mathrm{W\,cm^{-2}}$. The channels showed a quasi-parabolic radial electron density profile, which is able to guide more than 50 optical modes, and extended for the whole thickness of the gas-jet. It was proved that the nanosecond pulse generated such uniform long channel during its rise time. In fact, at this stage it is possible to find a condition in which the longitudinal plasma growth is dominated by the motion of the gas breakdown threshold rather than by the propagation of the blast wave [40, 41]. Figure 8.4 shows a typical interferogram and the retrieved transverse distribution of the plasma electron density, demonstrating the channeled shape for a plasma [41].

A further very promising technique relies on the production of plasma channels in gas-filled capillaries. In such devices, the ionization is induced via electric discharge through electrodes put at the edges of a thin tube containing low-Z gas. After the early experiments on thin glass capillaries, with the obvious drawback of single-shot operation due to capillary destruction after laser interaction but with promising results [42,43], capillaries made up of two half-channeled joint sapphire blocks filled with gas and with electrodes placed near each end have been successfully used to reach GeV electron energies [6]. When a capacitor discharges electric current through the electrodes, the gas is ionized and shaped as a hollow-channeled optical fiber able to guide the laser pulse focused at its entrance. The profile of the electron density for a forefront capillary device has been recently characterized by means of interferometric technique, showing also the possibility of tailoring the plasma profile to match efficient LWF acceleration issues [44].

Finally, it is worth mentioning that with the advent of ultra-high intensity laser systems, the possibility for an ultrashort laser pulse to be self-guided via

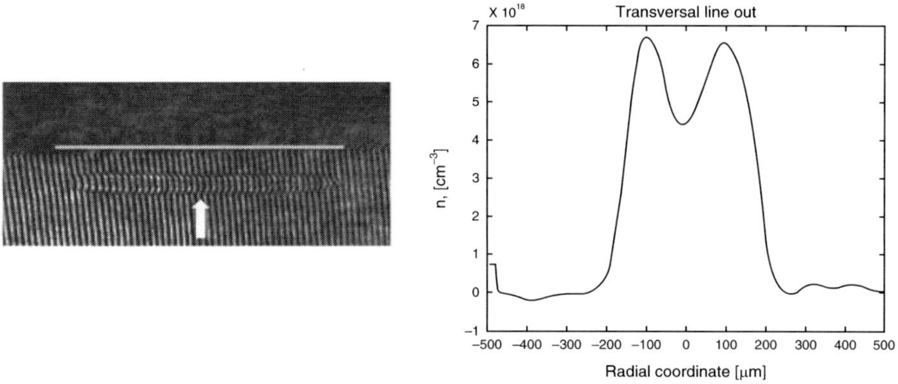

Fig. 8.4. Interferogram (*left*) and radial electron density distribution (*right*) for a plasma obtained with a ASE-like laser pulse, from [41]. The *arrow* and the *horizontal line* in the interferogram indicate the focus position and the 3 mm gas-jet slit position, respectively

interplay of relativistic self-focusing and ponderomotive self-channeling represents a further way to extend the propagation length at maximum intensity over several times Z_{R} [45, 46].

8.3 Electron Acceleration: Experiments and Simulations

8.3.1 Laser Wakefield Acceleration and Associated Regimes

About 20 years ago, the innovative laser technique of CPA [1] was well consolidated and made possible to experimentally investigate the laser wakefield acceleration (LWFA) concept first introduced by Tajima and Dawson as early as 1979 [47]. Currently operating CPA-based, table-top laser systems can provide pulses that, suitably focused, can reach peak intensity well above $10^{18}\,\mathrm{W\,cm^{-2}}$ and duration of tens of femtoseconds. The interaction of such a pulse with a plasma can excite electronic structures in its wake, displacing them to form an oscillating wave that can couple with the energy released by the pulse itself in its propagation [48, 49]. Large amplitude plasma waves can be then resonantly excited by the ponderomotive force of an ultrashort laser pulse, provided its duration τ matches the condition

$$c \cdot \tau \approx \lambda_{\mathrm{p}}/2, \tag{8.3}$$

where c is the light speed and $\lambda_{\mathrm{p}} = (\pi / r_{\mathrm{e}} n_{\mathrm{e}})^{1/2}$ is the plasma wavelength [47]. The electric field that is associated with plasma waves can be fruitfully exploited to accelerate electrons that can "surf" the excited waves in the wake of the propagating laser pulse. Further, unprecedented accelerating electric fields can be achieved, in principle exceeding $200\,\mathrm{GeV\,m^{-1}}$, almost four orders of magnitude higher than the ones available at ordinary, radiofrequency based, accelerators.

Several problems have been progressively faced by experimentalists in developing such a critical technique. The two main challenging questions to answer regard where the electrons to be accelerated can be taken from and how to manage to satisfy relation (8.3). Electron injection in plasma waves can be either external (e.g., an electron beam coming from an external linac can be focused in the plasma) or internal (electrons are taken directly from the breaking of the background plasma structures [50]). Because of the difficulty of synchronization for an external injection, acceleration experiments are mainly performed exploiting self-injection mechanisms. Available laser pulses hardly fulfill the relation (8.3) in terms of pulse duration (the high-density plasmas necessary for the self-trapping of electrons typically require sub 20 fs laser pulses for the relation to hold), but are so intense that they can excite in their wake a relativistic plasma wave via the onset of Raman forward scattering (RFS) parametric instability. In this scenario, the propagating electromagnetic wave decays into a forward propagating Stokes wave, an anti-Stokes wave, and a relativistic plasma wave. The beating of the laser electric field with the Stokes waves can eventually modulate the laser envelope, making it possible to be resonant with the plasma wave in the so-called self-modulated laser wakefield regime (SMLWF) [51]. In this framework, more than 10 years ago, first encouraging results initiated the path towards the optimization of a laser–plasma accelerator [52–54]. Further progresses of laser science and the availability of laser pulses with duration of few tens of femtoseconds and suitable power made then possible to enter quasi-resonance regimes closer to the theoretical requirement (8.3). This led to enter the forced laser wakefield (FLWF) scheme of acceleration, allowing electrons with energy up to 200 MeV to be produced with 30 fs, 2–3 J laser pulses at LOA laboratory in France [55] and the enhancement of the electron yield with 1 J, ~50 fs duration pulses at LBNL laboratory (USA) [56]. It has to be noted that most of these experiments have the common feature of poor quality of the electron beams. In the SMLWF regime, in fact, the laser pulse extends over several plasma periods, thus interacting with electron bunches accelerated by different plasma waves and degrading the output beam quality in terms of spatial collimation and monochromaticity. Even in the case of FLWF, the energy spread was found to reach values of approximately 100%, reducing the applicability of such high energy electron sources.

The first observation of monoenergetic beams from a laser–plasma accelerator rapidly followed the improved control of the plasma density, as well as the use of preformed hollow channels able to guide the laser pulse over several optical diffraction distances, and was reported by three groups from LOA, LBNL, and RAL laboratories [57–59] in 2004. Energies of about 100 MeV in acceleration paths of the order of 1 mm have been achieved in supersonic gas-jet targets, with energy spreads ranging from 2 to 24%. Theoretical knowledge of the underlying physics has also grown with the experimental results and laser science development, providing numerical proofs and supporting simulations for different regimes of laser–target interaction. The pioneering work of Pukhov and Mayer ter Vehn [46] showed that a highly nonlinear acceleration regime is onset when a sub-10 fs, 20 mJ laser pulse propagates in a plasma with $n_e = 3.5 \times 10^{19}\,\mathrm{cm}^{-3}$, or a 33 fs, 12 J pulse travels in a plasma with $n_e = 10^{19}\,\mathrm{cm}^{-3}$. In these two cases, 3D particle-in-cell numerical simulations show that the pulse propagates leaving a hollow sphere in its back, on the rear side of which a bunch of electrons can accumulate pushed by the radiation pressure of the electromagnetic pulse. This alternative mechanism of electrons loading can yield energies of 50 MeV in the first case and 300 MeV in the second case. In this framework, monoenergetic electrons production has been demonstrated [60, 61]. Extending the simulation parameters accounting for laser systems of hundreds of terawatts, new frontiers in LWF acceleration can be achieved. It has been shown that a 30 fs, 200 TW laser interacting with a 750 mm plasma with $n_e = 1.5 \times 10^{18}\,\mathrm{cm}^{-3}$ can produce high-charge (0.3 nC) electrons with energy of 1.5 GeV, while scaling laws predict that 120 GeV electrons can be generated with PW-class systems in low density plasma channels extending over tens of meters [62].

8.3.2 Advanced Techniques and Record Results

The laser–plasma acceleration process being very critical on both laser driver and plasma medium intrinsic characteristics, several effects can limit the energy gain in LWF acceleration process, such as optical diffraction and energy depletion of the pump laser pulse and/or slippage of the electrons with respect to the pushing plasma wave. Efficient acceleration takes place in distances that are upper-limited by the so-called *dephasing length* L_d:

$$L_d = \lambda_p^3/\lambda^2 \propto n_e^{-3/2}, \tag{8.4}$$

(λ is the laser wavelength) over which the relativistic electrons overtake the accelerating structure [63]. Pump depletion, too, limits the large amplitude wake waves excitation by the laser pulse [64]. As these effects strongly depend on n_e, target design and shaping is a major task in experimental campaigns. The control of n_e and of its spatial profile can provide a useful tool in order to maximize the electron energy E_e, reduce the energy spread $\Delta E_e/E_e$, increase the number of electrons in the bunch (bunch current), and extend the diffraction limited distance traveled by the focused pulse at high intensity. To achieve these improvements, several kinds of targets have been designed and tested. Among them, gas-jets, capillaries, and solid targets have up to now given important results.

Interaction of high-intensity laser pulse with supersonic gas-jets delivered by fast electromagnetic valves is the most common technique in LWF experiments worldwide. This kind of target allows a uniform, underdense plasma to be the interaction medium, with a density that can be tuned by managing the pressure of the neutral gas in the reservoir. Furthermore, the interaction length can be increased or reduced by changing the dimensions of the gas exit nozzle, while fast valves enable operation at a high repetition rate with a suitable synchronization with the laser pulse arrival. Injection of electrons in the plasma waves can moreover be improved by fine-tuning the laser interaction point with respect to the steep neutral density gradients at the gas entrance and exit edges. Gas-jets enable also a quite custom shaping, provided the preionization of the neutral gas by means of a subsidiary laser delivering an early pulse or of the amplified spontaneous emission (ASE) that precedes ultrashort pulses (see Sect. 8.2.3). Multipulse techniques enable to improve the electron beam quality. This latter is generally inversely measured by the "emittance" $\epsilon = \pi\sigma_r\sigma_\theta$, where σ_r and σ_θ are the radial and angular extension of the beam. High-quality electron bunches with $E_\mathrm{e} \approx 80\,\mathrm{MeV}$ and few π·mm·mrad emittance has been obtained with a multipulse technique: one first laser pulse ionizes a hydrogen plume while a second pulse heats and shapes the plasma where a third laser pulse creates the accelerating structure [65]. The same group has recently observed that focusing the laser pulse on the downramp of a hydrogen gas-jet can reduce the accelerated electron bunches emittance to tenth of π·mm·mrad [66]. Monoenergetic electron production with $E_\mathrm{e} \approx 200\,\mathrm{MeV}$ has been demonstrated in gas-jets, by optimizing the value of n_e to find a relation between λ_p and $c \cdot \tau$ in which the pulse duration is shorter than the plasma period and the plasma waves are bubble-shaped [60]. A technique of counter-propagating colliding laser pulses has been recently implemented to obtain a more efficient controlled injection of electron in a 2 mm helium gas-jet: it has been showed that with this technique electrons with E_e between 15 and 250 MeV, $\Delta E_\mathrm{e}/E_\mathrm{e} < 10\%$ and low divergence can be produced [67].

It must be said that, in terms of energy gain, a real progress to make plasma-based accelerators competitive with traditional RF-based facilities has been found in using gas-filled capillaries (see also Sect. 8.2.3). This technique enables at the same time to optically guide a propagating laser pulse and to have as low n_e on the longitudinal axis as required by L_d to be extended. With this kind of capillaries, developed by Hooker and coworkers at the Clarendon Laboratory of Oxford University, record electron energies of 0.5 and 1 GeV have been obtained with 12 and 40 TW laser pulses, respectively, in a 3.3 cm long capillary [6]. Guiding and high-energy gain acceleration has also been demonstrated in a 1.5 cm long sapphire capillary with 18 TW, 42 fs laser pulses, with reliable production of 200–300 MeV electrons and occasional achievement of 500 MeV [7].

A minor but promising line of investigation concerns the use of plasmas preformed from thin foil targets, via the exploding-foil technique. In this scheme, a thin foil of material, usually plastic of few micron thickness, is exploded in the interaction with the ASE pedestal. The main pulse interacts then with the preproduced plasma enabling the onset of the LWF mechanism. The choice of the thickness and material of the target, as well as

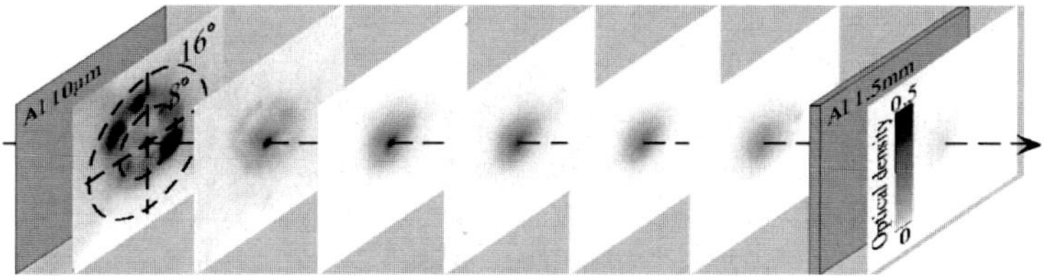

Fig. 8.5. Optical density spots produced by the released energy of electrons accelerated in exploding foils plasma (see also [68])

ASE optimization, must then account for the possibility of producing a suitable plasma profile for acceleration and for the produced electron bunches not to be attenuated in their path forward. In a pioneering experiment with thin plastic foils, ultracollimated bunches of electrons with E_e up to 40 MeV have been produced with a record angular divergence lower than 10^{-3} sr [68]. Figure 8.5 evidences that the electron angular distribution shows an outer ring of divergent, low energy electrons and an inner small spot of relativistic electrons.

Other experiments reported the production of high-energy electrons with a similar set-up, and showed the suitability of such relativistic bunches for nuclear activation techniques. It has been showed that nuclear reactions could be efficiently triggered by electrons from thin CH foil targets [69], and also that nuclear reactions can be a useful tool to characterize the electron beam accelerated with similar targets [70]. This point will be further developed in Sect. 8.4.

Recently, an interesting correlation between the laser pulse polarization and the ellipticity of the electron beam profile has been observed [71]. However, no major influence of laser polarization on the efficiency of the electron acceleration processes has been observed so far, nor this influence has been predicted by theory and simulations, differently from the proton acceleration. For proton acceleration, a great improvement on bunch charge and quality are expected by using circularly polarized laser pulses focused on thin foils at ultra-high intensities [72–74].

8.3.3 An Efficient Source of Relativistic Electrons for Medical Applications

In the previous paragraph, we have reported on some of the most exciting records of laser-driven accelerators competing with the existing accelerators devoted to high energy physics studies. However, relevant tasks have still to be pursued in this competition, like stability and reproducibility of the laser driving processes and the achievement of the required parameters of the electron bunches, including low-level emittance and high repetition rate. Nevertheless, if we limit our goal to a definite class of applications, as for some medical applications, present state of the art of table-top laser driven electron accelerators can be already suitable. In this context, most of the requirements on

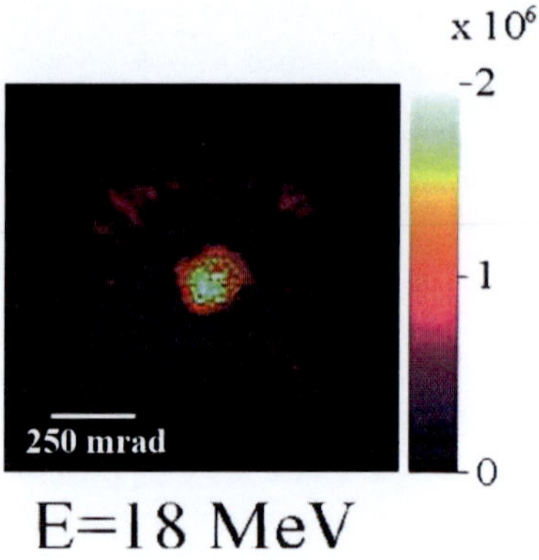

Fig. 8.6. Spatially resolved spectral data for 18 MeV accelerated electrons from the SHEEBA detector [75]

the electron bunches are considerably relaxed. Small divergence, monochromaticity, pointing stability, etc. are requested at a moderate level, while the main effort has to be devoted to the efficiency and reliability of the process.

In a recent experiment performed at CEA-Saclay (France), a regime of electron acceleration of high efficiency was found using a 10 TW laser and a supersonic jet of Helium [75]. A table-top accelerator was set, which delivered regularly high-charge (about 1 nC per Joule of laser energy), reproducible, fairly collimated, and quasi-monochromatic electron bunches, whose peak energy moved shot by shot in the range 10–45 MeV. The electron energy spectrum integrated over a large number of shots was found to peak at 21 MeV, with a FWHM energy spread of 8 MeV. As we will show below, this kind of integrated spectrum is suitable for a relevant class of cancer therapy.

Figure 8.6 shows a typical cross section of the 18 MeV component of the electron beam, after de-convolution of experimental data from the radiochromic film-stack device SHEEBA [76].

Three dimensional particle-in-cell simulation performed with the numerical code CALDER [77] reveals that the unprecedented efficiency of this accelerator was due to the achievement of a physical regime in which multiple electron bunches are accelerated in the gas-jet plasma during the action of each laser shot. This effect is shown in Fig. 8.7 by a snap-shot from the simulation sequence.

With this experiment, laser-driven electron acceleration approaches the stage of suitability for medical uses, in particular for Intra-Operative Radiation Therapy (IORT) of tumors [78, 79]. Comparison of the main parameters of electron bunches produced by two commercial RF Hospital accelerators for IORT treatment and those of this laser-driven accelerator is shown in the Table 8.1. Most of the performances of the experimental laser-driven electron accelerator set-up at CEA-Saclay are comparable with presently used conventional accelerators. Notice that, though the dose delivered for each shot is also comparable, the electron bunch duration is about six orders of

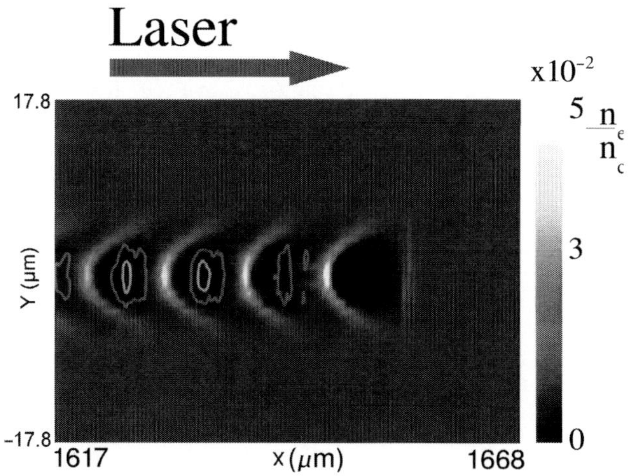

Fig. 8.7. Snapshot from the 3D simulation of the electron density distribution when the laser pulse is at about 1.65 mm after entrance in the gas jet

Table 8.1. Basic performances of two commercial medical linacs compared with the laser-driven electron accelerator

Accelerator type	IORT-NOVAC7	LIAC	Laser-driven accelerator
Producer	HITESYS SpA	SORDINA SpA	[75]
Electron energy	≤10 MeV	≤12 MeV	≥10 MeV
	(3, 5, 7, 9 MeV)	(4, 6, 9, 12 MeV)	(10–45 MeV)
Peak current	1.5 mA	1.5 mA	>1.6 kA
Bunch duration	4 μs	1.2 μs	<1 ps
Bunch charge	6 nC	1.8 nC	1.6 nC
Repetition rate	5 Hz	5–20 Hz	10 Hz
Mean current	30 nA @ 5 Hz	18 nA @ 10 Hz	16 nA @ 10 Hz
Delivered energy (1 min)	18 J @ 9 MeV	14 J @ 12 MeV	21 J @ 20 MeV

magnitude higher in the case of the laser-driven accelerator. Consequently, the peak current is 10^6 times higher. The biological and therapeutic effect of such huge difference is unknown. The new ultrashort laser-driven electron source thus opens also an exciting field of basic bio-medical research.

8.4 Nuclear Activation Using Electron Bunches from Laser Plasmas

Laser–plasma accelerators represent a unique tool for investigating matter properties at a reduced laboratory scale, compared to large-scale facilities. Nowadays, efficient electron acceleration can be provided by such table-top accelerators and can be used to perform experiments in a variety of fields.

Nuclear science in particular obtains from laser-driven electron sources a brand new input to perform interesting measurements in the context of many laboratories equipped with ultrashort powerful lasers. The ultrashort duration of these particle bunches represent a further attractive feature for these kinds of studies. In the following, we will focus on nuclear reaction induced by gamma radiation produced by bremsstrahlung of laser-produced electrons in suitable "radiator" targets. This way is usually mentioned as "photo-activation" and is particularly efficient for photons of energy close to the Giant Dipole Resonance of many nuclei.

8.4.1 Basics of Nuclear Photo-Activation

About 70 years ago, Bothe and Gentner, with a source of 17 MeV photons, for the first time found a cross section for photonuclear activation of various material samples two orders of magnitude greater than the one predicted by theory [80,81], attributing this to a resonance. Nearly 10 years later, Baldwin and Klaiber confirmed in photo-absorption experiments, such a strong resonance, that it was denoted to have electric dipole nature [82,83]. Eventually, it was found that this resonance is common to all nuclei, with variations (both in shape and width) depending on material atomic weight A; this behavior was addressed to collective proton vibrations towards neutrons as a restoring force applies when a particle passes nearby the nucleus. This Giant Dipole Resonance (GDR) lies in the range 10–30 MeV for most of the medium and heavy nuclei. The photo-absorption cross section assumes a Lorentzian shape:

$$\sigma(E) = \frac{\sigma_{\mathrm{m}}}{1 + [(E^2 - E_{\mathrm{m}}^2)^2 / E^2 \Gamma^2]} \tag{8.5}$$

where E_{m}, σ_{m}, and Γ are the resonance energy, maximum cross section, and FWHM, respectively [84].

In Table 8.2 the values of the key parameters for photo-absorption reactions are listed for a variety of elements usually employed in experimental tests.

The degree of activation of the sample is measured by post-irradiation spectroscopy, usually performed with high-purity semiconductors. The time-resolved intensity measurements of one of the several spectral lines enables to get the *half-life* of the radioactive element and the *total number* of nuclear reactions occurred. In fact, the intensity of a given spectral line associated with the decay of the radioactive elements decreases with time as $A(t) = A_0 \exp[-t/\tau]$, where A_0 indicates the initial number of nuclei (at $t = 0$) and τ is the decay time constant related to the element *half-life* ($\tau = \ln 2/t_{1/2}$), which can be measured. Integrating this relation from $t = 0$ to the total acquisition time, and weighting it with the detector efficiency and natural abundance lines, the *total number of reactions* N can be derived. Then, if one compares this number with the value obtained from the convolution of

Table 8.2. Threshold energy, cross section maximum, resonance energy, and FWHM for different photo-induced reactions

Target nucleus	Reaction	E_{thr}(MeV)	σ_{m}(mbarn)	E_{m}(MeV)	Γ(MeV)
^{12}C	(γ,n)	18.7	8.73	23.4	6
^{63}Cu	(γ,n)	10.8	78.6	16.6	6
	(γ,2n)	19.7	10	23.7	6.5
	(γ,3n)	31.4	–	–	–
^{65}Cu	(γ,n)	9.9	96	17	5
^{129}I	(γ,n)	8.8	–	–	–
^{181}Ta	(γ,n)	7.6	221	12.8	2.1
	(γ,3n)	22.1	21	27.7	5.6
^{197}Au	(γ,n)	8.1	529.2	13.5	4.5
	(γ,2n)	14.7	106.7	16.8	7
	(γ,3n)	23.1	13.6	27.1	¿6
^{232}Th	(γ,f)	6	54	14.2	7
^{238}U	(γ,f)	5.8	175	14.3	8.5

(γ,f) stands for photo-fission reaction. Data are retrieved from [85]

the reaction cross section and the gamma radiation spectrum (that has to be known with aid of other diagnostics):

$$N = N_0 \int_{E_{\mathrm{thr}}}^{\infty} \sigma(E) \cdot n_\gamma(E)\,\mathrm{d}E \tag{8.6}$$

where E_{thr} is the energy threshold for the reaction, $\sigma(E)$ is given by (8.5), and $n_\gamma(E)$ is the photon spectrum, the *spectrum of the γ-rays* responsible for the activation can be obtained.

8.4.2 Activation Induced by Laser–Plasma Electrons

In this context, electrons of tens of MeV that can be produced by laser–plasma accelerators are of particular interest. They have to be converted in high-energy photons to start nuclear processes. A converter is then needed, in which the electrons undergo bremsstrahlung radiation and produce a continuous gamma spectrum. Efficient conversion is found to take place for converter thickness from 0.3 to 0.5 times the mean range of the incident electrons [86], the most efficient one depending on electron energy. A widely used material for bremsstrahlung conversion is tantalum.

A laser-based activation experiment consists in concept of three stages: the first stage is the "accelerator", in which the laser, interacting with matter, produces relativistic electrons; this can be obtained either in laser–solid or in laser–gas interactions (see Sect. 10.3). The second stage, called in the following "radiator", converts the relativistic electrons in γ-rays via bremsstrahlung in a solid. The third stage, called in the following "activation sample" or simply "sample", consists of a target whose material is partially activated by the γ-rays.

In some experiments, the same solid target acted for both accelerator and radiator.

Let us first review some experiments in which the laser-driven acceleration of electrons has been obtained in laser–solid interactions.

Tantalum target has been used by Ledingham et al. with VULCAN laser at RAL [87] as accelerator and radiator, in which electrons were accelerated and generated high-energy photons. These latter successively impinged on a stack formed by different materials. At an intensity of $10^{19}\,\mathrm{W\,cm^{-2}}$, the ($\gamma$,n) reactions have been investigated for ^{12}C, ^{39}K, ^{63,65}Cu, ^{64}Zn, ^{107}Ag, ^{141}Pr, and ^{181}Ta, revealing activities of about 5 kBq, and 6×10^4 laser generated nuclear fissions of ^{238}U per Joule of laser energy. In another experiment performed at LLNL with NOVA laser operating at $10^{20}\,\mathrm{W\,cm^{-2}}$, using a slab of gold as accelerator and radiator, Cowan et al. found approximately the same fission rate in uranium [88].

Schwoerer et al. used tantalum for both accelerator and radiator but in two separate targets and the Jena 15 TW laser system at an intensity of $10^{20}\,\mathrm{W\,cm^{-2}}$. A 50 µm first foil employed for the accelerator and a second slab of 1 mm that acted as radiator [89]. As a result, 10^4 fission events per Joule of laser energy of ^{232}Th and ^{238}U, placed behind the second tantalum layer, were obtained with a reaction rate of the order of 1 event per laser shot.

Even transmutation of elements that are of environmental risk, like ^{129}I, have been investigated: for example, ^{128}I was successfully obtained from ^{129}I by Magill et al. in 2003 and by Ledingham et al. in the same year [90, 91]. In the first case, the Jena laser delivering 1 J at intensities up to $10^{20}\,\mathrm{W\,cm^{-2}}$ was focused on a 2 mm tantalum sheet (accelerator and radiator): photoneutron reactions were found to be 2 per laser shot. In the latter case, a single VULCAN laser pulse of 360 J, focused at $5 \times 10^{20}\,\mathrm{W\,cm^{-2}}$ on a 4 mm thick gold target, generated about 3×10^6 nuclei of ^{128}I. Finally, Spencer et al. used the ultra-high intensity of VULCAN laser to develop a photonuclear reaction-based laser–plasma diagnostic [92], while Malka et al. reported about the characterization of high-energy electron bunches generated in laser–solid interactions by means of a spectrometer and photonuclear reactions for a 30 TW laser system at Laboratoire d'Optique Appliquée (France) [69].

A critical point in the retrieving of the number of nuclear reactions in laser–solid experiments is that there is no control on the spectrum of the electrons accelerated in the interaction, as well as the acceleration mechanism is uncertain and difficult to fit in a predictable scheme. In most cases, the electron energy distribution is assumed to be Boltzmann-like and deconvolutions are performed starting from this assumption.

A better control on the electron spectrum can be obtained in experiments that employ gas-jets as target for laser interaction. Experiments performed with gas-jets enable the generation of electron bunches with a higher energy and a lower angular spread.

Among the experiments of nuclear activation driven by LWF accelerated electrons, Leemans et al. in 2001 performed a scan of the interaction conditions of l'OASIS laser pulse with gas-jets of H_2, He, and N_2 by changing laser pulse duration and delivered energy as well as gas backing pressures and position of the laser focus with respect to the gas flow [93]. They found that for laser power of 8.3 TW, pulse duration of 50 fs, and plasma densities of $1–3 \times 10^{19}\,\mathrm{cm}^{-3}$, the accelerated electrons could efficiently generate bremsstrahlung photons in a Pb radiator and induce the reactions $^{63}\mathrm{Cu}(\gamma,\mathrm{xn})^{63-x}\mathrm{Cu}$, with $x = 1, 2$. Liesfeld et al. have been able, focusing the Jena laser at a nominal intensity of $5 \times 10^{19}\,\mathrm{W\,cm}^{-2}$ in a $5\,\mu\mathrm{m}^2$ spot size onto a helium gas-jet, to improve their earlier results of [90] on the transmutation of ^{129}I of more than two orders of magnitude: $2.16 \cdot 10^5$ reactions with 10^4 laser shots [94]. Furthermore, their electron beams efficiently activated photoneutron reactions in ^{181}Ta and ^{197}Au that were used as radiator and sample. Finally, it is interesting to note how laser–gas jet production of electron can enhance the number of photo-induced fissions in ^{238}U of roughly one order of magnitude with respect to experiments on solid targets cited above: this has been recently proved by Reed et al. [95] by firing only 72 times a 30 fs pulse with intensity of $10^{19}\,\mathrm{W\,cm}^{-2}$ on a 2 mm He jet at $3 \times 10^{19}\,\mathrm{cm}^{-3}$ plasma electron densities. A number of ^{238}U fissions as high as 3×10^5 was obtained by means of careful optimization of the electron beam spatial profile, total charge, and energy by varying nozzle position and gas density.

8.4.3 An Example of High-Efficiency Photonuclear Activation in a Gas-Jet Experiment

Recently it has been demonstrated that, with a table-top 10 TW laser system, it is possible to set-up a high-efficiency, high-energy electron source to be used for nuclear study purposes. The laser system dedicated to this goal was the one mentioned in Sect. 8.3.3, while the set-up of the experiment is slightly different, with the activating stuff (radiator + sample) placed after the gas-jet instead of the spectrometers. The investigated reaction is $^{197}\mathrm{Au}(\gamma,\mathrm{n})^{196}\mathrm{Au}$, induced by photons generated by electrons bremsstrahlung in a 2 mm tantalum slab. The layout of the experiment is schematically shown in Fig. 8.8(top-left).

As it is shown in Fig. 8.8 and as it is described in Sect. 8.3.3, the accelerated electron energies fit very well in the GDR range for the chosen photo-induced reaction. 106 laser shots produced an important amount of photo-activations, as it was detected by a high-purity germanium detector. The sample activity after irradiation was measured considering the number of counts in the germanium detector under each energy peak. The trend of the counts number with post-irradiation measurement time gave also the estimate of the ^{196}Au half-life, accounting for 6.09 ± 0.26 days (in good agreement with the tabulated 6.17 value [85]). The calculation performed with the Montecarlo code GEANT4 [96], accounting for the experimental geometry, the electron spectrum, and the photonuclear cross section, gives an absolute reaction rate of

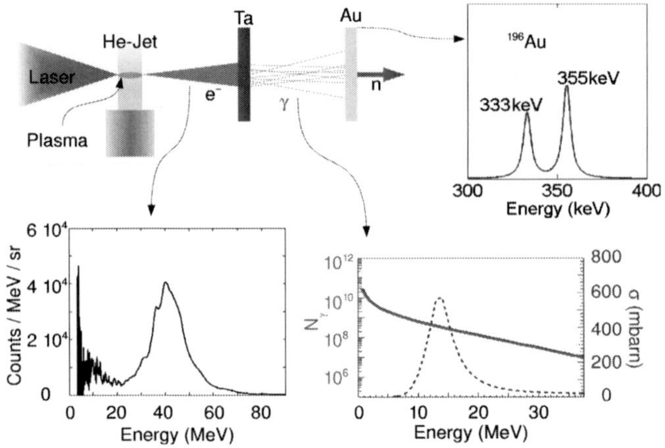

Fig. 8.8. Experiment set-up and obtained results. Counter-clockwise, from top left: schematic view of the set-up for the activation experiment, accelerated electron spectrum as measured by a magnetic spectrometer, full differential cross section for the reaction $^{197}\mathrm{Au}(\gamma, \mathrm{n})^{196}\mathrm{Au}$ (*dotted*) and bremsstrahlung spectrum reconstructed after the deconvolution calculations (*solid*), x-ray emission spectrum of $^{196}\mathrm{Au}$ measured after the gold sample irradiation

1.46×10^6 per Joule of incident laser energy. This large value, achieved with a relatively small number of laser shots, evidences the quality of the electron beam produced in the experiment; the resulting number of electrons with energy greater than 3.4 MeV was found to be $(3.15 \pm 0.13) \times 10^{10}$ per laser shot [75].

8.4.4 Perspectives of Possible Applications

Until this rapid evolution of laser-triggered nuclear reactions, the only possibility was to use large accelerators for experimental tests. The recent results open to the possibility of performing experiments and a variety of applications in many fields of nuclear interest by exploiting tools that are now available in small-scale facilities and on day by day basis [97].

Concerning basic nuclear science studies, it is worth citing the possibility of performing measurements of nuclear data, such as photonuclear cross sections on radioisotopes, or the investigation of the fission in actinide elements, or the transmutation of fission products. Furthermore, it could be possible to measure the polarization of the laser-produced electron bunches by using the transmission polarimetry technique. Among further uses that can be foreseen, laser triggered transmutation reactions could provide isotopes for biomedical use, and reduce the environmental risk represented by the presence of long-lived fission products (like the $^{129}\mathrm{I}$ mentioned in the text), even if the elimination of nuclear waste would require reaction rates that are much higher than the ones available nowadays.

8.5 Conclusion

The use of ultrashort intense laser pulses led to an impressive progress in the domain of high field photonics in the last two decades. Basic knowledge on laser pulse interaction with ionized media in a large variety of conditions became available. This allowed to control the propagation of the laser pulse in a suitable plasma to optimize the processes involved in the excitation of plasma waves, capture and acceleration of free electrons up to relativistic kinetic energies. In particular, millimetre sized electron accelerators driven by table-top lasers have been recently set-up, which are suitable for biomedical applications. The relativistic electron bunches delivered by such accelerators have in turn generated, via bremsstrahlung, gamma-rays with an intense component in the Giant Dipole Resonance of nuclei. These gamma-ray sources have been successfully used for tests of photonuclear activation. Our review, far from being exhaustive, already suggests that research in high field photonics opened an exciting broad perspective of advanced knowledge and practical applications.

Acknowledgements

Authors are grateful to Philippe Martin and all the colleagues of the PHI group from CEA-Saclay; to Danilo Giulietti, Leonida A. Gizzi, Luca Labate, and all the other colleagues of the ILIL team from CNR, INFN, and University of Pisa.

References

1. D. Strickland, G. Mourou, Opt. Comm. **56**, 219 (1985)
2. A. Pukhov, Rep. Prog. Phys. **66**, 47 (2003)
3. T. Brabec, F. Krausz, Rev. Mod. Phys. **72**, 545 (2000)
4. D.Giulietti, L.A. Gizzi, A. Giulietti, et al., Phys. Rev. Lett. **79**, 3194 (1997)
5. D. Umstadter, J. Phys. D Appl. Phys. **36**, R151 (2003)
6. W.P. Leemans, B. Nagler, A. Gonsalves, et al., Nat. Phys. **2**, 696 (2006)
7. S. Karsch, J. Osterhoff, A. Popp, et al., New J. Phys. **9**, 415 (2007)
8. A. Maksimchuk, S. Reed, S.V. Bulanov, et al., Phys. Plasma **15**, 056703 (2008)
9. H. Schwoerer, S. Pfotenhauer, O.Jäckel, et al., Nature **439**, 445 (2006)
10. B.M. Hegelich, B.J. Albright, J. Cobble, et al., Nature **439**, 441 (2006)
11. G. Mourou, T. Tajma, S.V. Bulanov, Rev. Mod. Phys. **78**, 309 (2006)
12. S. Miyamoto, J. Plasma Fusion Res. **73**, 688 (1997)
13. P. Gibbon, *Short Pulse Laser Interactions with Matter* (Imperial College Press, London, 2005)
14. D. Teychenné, A. Giulietti, D. Giulietti, et al., Phys. Rev. E **58**, R1245 (1998)
15. W. Yu, Z.M. Sheng, M.Y. Yu, et al., Phys. Rev. E **59**, 3583 (1999)
16. R.A. Cairns, B. Rau, M. Airila, Phys. Plasma **7**, 3736 (2000)
17. G. Ferrante, M. Zarcone, S.A. Uryupin, Phys. Rev. Lett. **91**, 085005 (2003)

18. V.A. Vshivkov, N.M. Naumova, F. Pegoraro, et al., Nucl. Instrum. Meth. A **410**, 493 (1998)
19. V.A. Vshivkov, N.M. Naumova, F. Pegoraro, et al., Phys. Plasma **5**, 2727 (1998)
20. J. Fuchs, J.C. Adam, F. Amiranoff, et al., Phys. Rev. Lett. **80**, 2326 (1998)
21. J. Dawson, Rev. Mod. Phys. **55**, 403–447 (1983)
22. S.C. Wilks, J.M. Dawson, W.B. Mori, Phys. Rev. Lett. **61**, 337 (1988)
23. V. Popov, Phys. Usp. **47**, 855 (2004)
24. V. Popov, J. Exp. Theor. Phys. **91**, 48 (2000)
25. M.V. Ammosov, N.B. Delone, V.P. Krainov, Sov. Phys. JETP **64**, 1191 (1986)
26. A. Giulietti, P. Tomassini, M. Galimberti, et al., Phys. Plasma **13** 093103 (2006)
27. P. D'Oliveira, S. Dobosz, S. Hulin, et al., J. Opt. Soc. Am. B **19**, 2603 (2002)
28. S. Dobosz, P. D'Oliveira, S. Hulin, et al., Phys. Rev. E **65**, 047403 (2002)
29. P. Tomassini, A. Giulietti, L.A. Gizzi, et al., Appl. Opt. **40**, 6561 (2001)
30. K. Nugent, Appl. Opt. **24**, 3101 (1985)
31. V. Malka, C. Coulaud, J.P. Geindre, et al., Rev. Sci. Instrum. **71**, 2329 (2000)
32. P. Tomassini, A. Giulietti, Opt. Comm. **199**, 143 (2001)
33. P. Tomassini, A. Giulietti L.A. Gizzi, et al., Laser Part. Beams **20**, 195 (2002)
34. A. Giulietti, M. Galimberti, A. Gamucci, et al., Laser Part. Beams **25**, 513 (2007)
35. T. Hosokai, K. Kinoshita, A. Zhidkov, et al., Phys. Rev. Lett. **97**, 075004 (2006)
36. P. Sprangle, E. Esarey, J. Krall, et al., Phys. Rev. Lett. **69**, 2201 (1992)
37. C.G. Durfee III, H.M. Milchberg, Phys. Rev. Lett. **71**, 2409 (1993)
38. E. Esarey, P. Sprangle, J. Krall, et al., IEEE J. Quant. Electron. **33**, 1879 (1997)
39. P. Volfbeyn, E. Esarey, W.P. Leemans, Phys. Plasma **6**, 2269 (1999)
40. D. Giulietti, A. Giulietti, M. Lucchesi, et al., J. Appl. Phys. **58**, 2916 (1985)
41. A. Gamucci, M. Galimberti, A. Giulietti, Appl. Phys. B **85**, 611 (2006)
42. F. Amiranoff, S. Baton, D. Bernard, et al., Phys. Rev. Lett. **81**, 995 (1998)
43. B. Cros, C. Courtois, G. Malka, et al., IEEE Trans. Plasma Sci. **28**, 1071 (2000)
44. A. Gonsalves, T. Rowlands-Rees, B. Broks, et al., Phys. Rev. Lett. **98**, 025002 (2007)
45. W.P. Leemans, E. Esarey, C. Schroeder, et al., Phil. Trans. R. Soc. A **364**, 585 (2006)
46. A. Pukhov, J. Meyer-Ter-Vehn, Appl. Phys. B **74**, 355 (2002)
47. T. Tajima, J. Dawson, Phys. Rev. Lett. **43**, 267 (1979)
48. J.R. Marquès, J.P. Geindre, F. Amiranoff, et al., Phys. Rev. Lett. **76**, 3566 (1996)
49. N. Matlis, S. Reed, S.V. Bulanov, et al., Nat. Phys. **2**, 749 (2006)
50. A.I. Akhiezer, R.V. Polovin, et al., Sov. Phys. JETP **3**, 696 (1956)
51. C. Joshi, et al., Phys. Plasma **14**, 055501 (2007)
52. A. Modena, Z. Najmudin, A.E. Dangor, et al., Nature **377**, 606 (1995)
53. K. Nakajima, D. Fisher, T. Kawakubo, et al., Phys. Rev. Lett. **74**, 4428 (1995)
54. F. Amiranoff, S. Baton, D. Bernard, et al., Phys. Rev. Lett. **81**, 995 (1998)
55. V. Malka, S. Fritzler, E. Lefebvre, et al., Science **298**, 1596 (2002)
56. W.P. Leemans, P. Catravas, E. Esarey, et al., Phys. Rev. Lett. **89**, 174802 (2002)
57. J. Faure, Y. Glinec, A. Pukhov, et al., Nature **431**, 541 (2004)
58. C.G.R. Geddes, C. Toth, J. van Tilborg, et al., Nature **431**, 538 (2004)
59. S.P.D. Mangles, C. Murphy, Z. Najmudin, et al., Nature **431**, 535 (2004)
60. V. Malka, J. Faure, Y. Glinec, et al., Phys. Plasma **12**, 056702 (2005)
61. B. Hidding, K.U. Amthor, B. Liesfeld, et al., Phys. Rev. Lett. **96**, 105004 (2006)

62. W. Lu, M. Tzoufras, C. Joshi, et al., Phys. Rev. ST Accel. Beams **10**, 061301 (2007)
63. E. Esarey, P. Sprangle, J. Krall, et al., IEEE Trans. Plasma Sci. **24**, 252 (1996)
64. S.V. Bulanov, I.N. Inovenkov, V.I. Kirsanov, et al., Phys. Fluid B **4**, 1935 (1992)
65. C.G.R. Geddes, C. Toth, J. van Tilborg, et al., Phys. Rev. Lett. **95**, 145002 (2005)
66. C.G.R. Geddes, K. Nakamura, G.R. Plateau, et al., Phys. Rev. Lett. **100**, 215004 (2008)
67. J. Faure,C. Rechatin, A. Norlin, et al., Nature **444**, 737 (2006)
68. D. Giulietti, M. Galimberti, A. Giulietti, et al., Phys. Plasma **9**, 3655 (2002)
69. G. Malka, M. Aleonard, J. Chemin, et al., Phys. Rev. E **66**, 066402 (2002)
70. M. Gerbaux, F. Gobet, M. Aleonard, et al., Rev. Sci. Instrum. **79**, 023504 (2008)
71. S.P.D. Mangles, A.G.R. Thomas, M.C. Kaluza, et al., Phys. Rev. Lett. **96**, 215001 (2006)
72. A. Macchi, F. Cattani, T.V. Liseykina, et al., Phys. Rev. Lett. **96**, 165003 (2005)
73. X.Q. Yan, C. Lin, Z.M. Sheng, et al., Phys. Rev. Lett. **100**, 135003 (2008)
74. Y.I. Salamin, Z. Harman, C.H. Keitel, Phys. Rev. Lett. **100**, 155004 (2008)
75. A. Giulietti, N. Bourgeois, T. Ceccotti, et al., Phys. Rev. Lett. **101**, 105002 (2008)
76. M. Galimberti, A. Giulietti, D. Giulietti, et al., Rev. Sci. Instrum. **76**, 053303 (2005)
77. E. Lefebvre, N. Cochet, S. Fritzler, et al., Nucl. Fusion **43**, 629 (2003)
78. U. Veronesi, G. Gatti, A. Luini, et al., Breast J. **9**, 106 (2003)
79. A.S. Beddar, P.J. Biggs, S. Chang, et al., Med. Phys. **33**, 1476 (2006)
80. W. Bothe, W. Gentner, Z. Phys. **106**, 236 (1937)
81. A.B. Migdal, J. Phys. USSR **8**, 331 (1944)
82. G.C. Baldwin, G.S. Klaiber, Phys. Rev. **71**, 3 (1947)
83. G.C. Baldwin, G.S. Klaiber, Phys. Rev. **73**, 1156 (1948)
84. J. Speth, A. Van Der Woude, Rep. Prog. Phys. **44**, 719 (1981)
85. IAEA Photonuclear Data Library, online at http://www-nds.iaea.org/photonuclear/
86. M.J. Berger, S.M. Seltzer, Phys. Rev. C **2**, 621 (1970)
87. K.W.D. Ledingham, I. Spencer, T. McCanny, et al., Phys. Rev. Lett. **84**, 899 (2000)
88. T.E. Cowan, A. Hunt, T. Phillips, et al., Phys. Rev. Lett. **84**, 903 (2000)
89. H. Schwoerer, F. Ewald, R. Sauerbrey, et al., Europhys. Lett. **61**, 47 (2003)
90. J. Magill, H. Schwoerer, F. Ewald, et al., Appl. Phys. B **77**, 387 (2003)
91. K.W.D. Ledingham, J. Magill, P. McKenna, et al., J. Phys. D Appl. Phys. **36**, L79 (2003)
92. I. Spencer, K.W.D. Ledingham, R. Singhal, et al., Rev. Sci. Instrum. **73**, 3801 (2002)
93. W.P. Leemans, D. Rodgers, P. Catravas, et al., Phys. Plasma **8**, 2510 (2001)
94. B. Liesfeld, K.U. Amthor, F. Ewald, et al., Appl. Phys. B **79**, 1047 (2004)
95. S. Reed, V. Chvykov, G. Kalintchenko, et al., Appl. Phys. Lett. **89**, 231107 (2006)
96. S. Agostinelli, et al., Nucl. Instrum. Meth. Phys. Res. A **506**, 250 (2003)
97. J. Galy, M. Maucec, D. Hamilton, et al., New J. Phys. **9**, 23 (2007)

9

Laser Plasma Acceleration and Related Electromagnetic Sources

Danilo Giulietti and Luca Labate

Summary. In this Chapter a brief overview will be given of those aspects of Laser Plasma Acceleration leading to a completely new generation of compact sources of energetic particles and photons to be foreseen. In particular, a simplified introduction to the Field of electron Laser Wakefield Acceleration will be first given and some recent experiments described. The emerging issue of protons acceleration will be then briefly discussed. Finally, an overview will be given of the recently studied and experimentally discovered processes leading to the production of ultrashort X-ray pulses via the interaction of ultraintense optical laser beams.

9.1 Introduction

In the last 1970s, Tajima and Dawson [1] proposed an innovative particle acceleration scheme, the so-called *laser wakefield acceleration* (LWFA) based on the laser pulse propagation in a plasma. The laser pulse, via the ponderomotive force (proportional to the gradient of the laser intensity) acting on the electron density, induces huge electric fields ($100\,\mathrm{GV\,m^{-1}}$ and greater) up to four orders of magnitude greater than the ones set in the conventional accelerators. The availability of such enormous fields (consequence of the electrons–ions separation) is basically due to the matter being already ionized in a plasma, so that no limitation is needed on the field strength to avoid the material breakdown. The electrons expelled by the most intense laser pulse region accumulate just behind the e.m. packet, thus producing an electron density perturbation in the wake of the laser pulse itself. Depending on the intensity and other laser pulse parameters (like the longitudinal and transversal dimensions), such a perturbation can assume different shapes and plasma waves in a nonlinear regime can be excited.

The physical situation is very similar to that of a boat moving on the sea surface, the gravity playing the role of the ponderomotive force in the previous case. The water moves from the immersed volume of the boat towards the stern, where it accumulates in the first crest of a wave produced in the boat wake. The amplitude of this wave depends on the volume of the immersed part of the boat and other parameters like the shape and the velocity of the boat. However, there is a major difference between the two cases, related to the nature of the two media in which such similar phenomena develop. Although the water is a nonresonant medium, a plasma exhibits a resonant frequency, the so-called *plasma frequency* $\omega_p = \sqrt{4\pi n_e e^2/m_e}$, at which any electron density perturbation evolves in time. It follows that the excitation of the plasma wave is strongly enhanced when the laser pulse length fits in half of the plasma wavelength $\lambda_p \approx 2\pi c/\omega_p$ (*Langmuir length*). In this case, the ponderomotive force associated with the laser pulse expels electrons from a region whose dimension matches the one of a density minimum region of a plasma wave. With this respect, this wakefield generation process can be considered as *quasi-resonant.*

Once excited, the plasma wave has the right characteristics to accelerate electrons that are injected in it up to very high energy. Indeed, in the usual experimental conditions, the huge longitudinal electric field moves with a phase velocity, set by the laser pulse group velocity, very close to c. However, to gain as much energy as possible from the plasma wave, electrons need not only to be injected into the plasma wave with the suitable phase, but also to experience an electric field of the same sign during the whole acceleration process. To be fulfilled, this condition requires, from the beginning of the process, relativistic electrons with an initial velocity close to c, which then increase their energy during the acceleration process. Once again this physical situation is very similar to another one that involves the sea waves and the surfers. Indeed, the surfer goes into the wave in a suitable phase of the wave and with an appropriate velocity. Only in this way can he acquire further velocity, while he is descending from the crest to the valley of the wave. After that he has to leave the wave, otherwise he is suddenly stopped.

The maximum energy gained by the electron in this acceleration process depends essentially upon the amplitude of the accelerating electric field, that is of the excited plasma wave, and the length of the path along which this acceleration occurs. Although a general approach to this issue is very complicated, in particular when nonlinear processes have to be considered, a 1D model [2, 3] in the case of small amplitude plasma waves can provide a satisfactory description, at least from a qualitative point of view, of the physical situation. In the frame of this model, if the velocity of the injected electron is already relativistic, the dephasing length comes out to be (see as an example [4] and therein) $L_{\mathrm{deph}} \approx \gamma_p^2 \lambda_p$, where $\gamma_p = \omega_L/\omega_p$ is the Lorentz factor of the plasma wave (ω_L is the laser angular frequency). We can thus see from these simple arguments that, while the maximum accelerating electric field scales as $n_e^{1/2}$ (as a consequence of the largest charge separation attainable in a plasma with a density n_e), the dephasing length scales in turn as $n_e^{-3/2}$ (as a consequence of the highest phase velocity of the plasma wave, imposed by the laser pulse group velocity). This leads to the maximum attainable energy scaling as $1/n_e$.

From an experimental viewpoint, the situation is much more complicated. Indeed, when the plasma density is decreased so that the acceleration length can reach the centimeter scale, actually we face with the difficulty of creating so long and homogeneous laser plasmas. Moreover, to create and maintain high amplitude plasma waves over such long distances, the laser intensity, on which the ponderomotive force depends, needs also to be kept sufficiently high. This condition is very difficult to be fulfilled due to the fundamental laws governing the focusing of an e.m. beam. As it is well known, in the case of a gaussian beam propagating in vacuum, the longitudinal extent of the most intense region of the beam is quadratically related to the transversal one, as shown by the expression of the *Rayleigh length*, the distance, along the propagation direction, from the waist to the point where the intensity decreases by a factor 2: $z_{\mathrm{R}} = \pi w_0^2/\lambda$, w_0 being the beam size in the waist. According to this formula, the intensities required for the wakefield acceleration scheme can be only maintained, with the laser pulse energies currently available, on a sub-millimeter scale. However, this length can be increased, for example, by suitably preforming a plasma channel and/or exploiting such phenomena as relativistic and ponderomotive self-focusing (see, e.g., [5,6]). Moreover, a very promising technique to overcome the above mentioned difficulties is to let the laser pulse propagate in a suitable hollow optical fiber in which a preformed plasma is created via electrical discharge. An outstanding result has been recently reported using this technique, which allowed to guide the laser over a $\approx$10 cm length and to get electrons accelerated up to 1 GeV [7].

At this point a question might arise: why we had to wait the advent of high intensity, ultrashort laser pulses for developing the LWFA technique? The reasons are essentially two. The first and simpler one is that the high intensities required for the generation of electron plasma waves are easily achievable only with such lasers. However, the main reason is that, because of the "quasi- resonant" nature of the process responsible for the laser wakefield process to occur, electron plasma waves at relatively high densities (10^{18}–$10^{19}\,\mathrm{cm}^{-3}$), required for obtaining an extremely high accelerating electric field (see above), can only be excited with sub 100 fs laser pulses.

The huge traveling electric field produced by the charge separation in an electron plasma waves is not able to accelerate the ions of the plasma, due to their much larger inertia with respect to the electrons. However, a different process has been discovered and studied in the field of laser-produced plasmas, which leads to the production of energetic protons/ions. Indeed, by focusing the laser on a thin solid target, suprathermal electrons with energy up to some MeV can be generated. These electrons, when leaving the solid/plasma target, generate a huge static electric field, due to the charge separation between electrons and ions, up to 1 TeV m^{-1} or greater, which is then able to accelerate the protons/ions up to some tens of MeV [8].

Besides these noteworthy features for the acceleration of electrons and ions, the use of powerful, ultrashort laser pulses has also opened the possibility of creating new X/γ-ray sources of unprecedented brightness and shortness.

The most direct and easy way consists in focusing the laser pulse onto a solid target and to collect the radiation emitted by the produced plasma. The wide emitted spectrum extends from infrared to X-rays and it is produced by different physical mechanisms: Bremsstrahlung, recombination, resonant lines, K-shell emission from neutral (or partially ionized) atoms. In particular, this latter mechanism has been recognized, since a decade, as a way of producing ultrashort monochromatic radiation pulses at energy up to several keV.

A different, very promising scheme to produce monochromatic X-ray radiation consists in letting an electron bunch interacting, possibly head-on, with an intense laser pulse, thus producing, via Thomson Scattering, an X-ray pulse. If the electron bunch is monochromatic, the monochromaticity of the X-ray pulse basically depends on the acceptance angle of the collected radiation. In this way, monochromatic X-ray beams can be produced, whose photon energy can be possibly tuned by changing the energy of the electron bunch. In this case, the required experimental setup consists in a 10–100 MeV LINAC and a pulsed laser delivering several joules per pulse, possibly at high repetition rate. An alternative, more compact setup, which can now be foreseen as the results discussed above are considered, relies only on the use of a powerful ultrashort laser pulse. The same laser system can be used, after having been splitted, for the production of the energetic electron bunch via LWFA in a plasma produced by a part of the beam as well as for the X-ray production via Thomson scattering of the remaining part by the electron bunch. The advantages of this scheme, in terms of smaller costs and reduced size, are evident. Of course, in this case the spectrum of the X-ray beam will be much wider, as it depends on the energy spread of the electron bunch, which is currently limited to about 10%.

A completely different emission process, which can in principle provide table-top ultrashort X-ray sources up to 100 keV has been recently discovered and studied, both from an experimental and a theoretical viewpoint [9]. It can be understood as one consider that the electrons, trapped and accelerated in a plasma wake as described earlier, can also experience, in some cases, a transverse force pulling them toward the beam axis. This force is basically due to the creation of a sort of plasma channel at low electron density, which is a consequence of the ponderomotive force that expels the electrons from the laser beam axis (the ions, due to their larger inertia, being fixed). The trapped electrons thus undergo a sort of wiggler motion, thus producing so-called betatron radiation.

Finally, we mention here a recent experiment in which γ-rays were produced, in a quite stable and efficient way, by letting relativistic electrons generated by laser wakefield acceleration impinge on a solid massive target [10]. These electrons then produced, via Bremsstrahlung emission, γ-ray radiation in the so-called Giant Dipole Resonance region, where efficient activation of materials can occur. We do not go into further details here, as this will be done in a different chapter by Giulietti et al.

The energetic particles produced by LPA and the e.m. radiation that such particles can generate by different processes allow several applications in different fields and particularly in the medical one. The achievement of ultra-high intensity X-ray radiation beams will enable to create new plasma states of astrophysical or geophysical interest. The extremely short duration of the energetic particle bunches produced, intrinsically synchronized with the ultrashort laser pulses, opens to unprecedented experiments in which particle and photons can be used in pump-and-probe experiments. In what follows, a brief description of some recent experimental results on the above issues will be given. In particular, Sect. 9.2 will focus on electron acceleration and Sect. 9.3 on protons acceleration. Finally, high energy photon sources based on laser-plasmas will be dealt with in Sect. 9.4.

9.2 Relativistic Electrons Sources

The first LPA experiments were based on the excitation of plasma waves induced by the beats of two long laser pulses, with angular frequency ω_1 and ω_2, respectively, propagating in a preformed plasma satisfying the resonant condition $\omega_p = \omega_1 - \omega_2$ (*beat wave* scheme). In this way some pioneering results have been obtained, with electrons accelerated to the 1–30 MeV range [11,12]. Although having been very important to demonstrate that the Tajima and Dawson idea could practically work, these experiments evidenced at the same time the difficulty, for example, due to the onset of instabilities, of maintaining very well defined laser and plasma conditions along a path long enough. As an example, the plasma density had to be kept fixed within less than 1% in order for the resonant condition to be fulfilled. For this reason, the technique was soon abandoned. Nearly at the same time, a different scheme now called *self-modulated wakefield accelerator* has been demonstrated to be able to accelerate electrons up to energy of some tens of MeV by using a relatively long laser pulse [13–15]. For the sake of brevity, we do not consider this scheme here.

After these experiments, in the mid-1990s, the overwhelming development of table-top multi TW, ultrashort pulse TiSa lasers, based on the CPA technique, suddenly changed the scenario. Indeed, in laser–plasma experiments at high intensity, mainly conceived to investigate the physical mechanisms relevant for the Inertial Confinement Fusion (ICF), more and more energetic electrons were produced as the laser intensities increased and the pulse duration decreased. As briefly explained in the Introduction, both these laser parameters are crucial for the electron plasma wave excitation: the intensity, proportional to the ponderomotive force, and the short pulse duration, which makes possible the quasi-resonance plasma wave excitation at high plasma density. Reducing the laser pulse duration while still increasing its intensity led to the so-called *laser wakefield acceleration* (LWFA) scheme to be directly accessed with table-top laser systems. As briefly explained in the introduction, within this scheme a short laser pulse directly creates a blowout region of depleted electron density due to the ponderomotive force $\mathbf{f}_{\mathrm{pond}} = -(e^2/4m_e\omega^2)\nabla E_0^2$ (in the nonrelativistic limit).

As an example of the number of experiments carried out in the last few years on this topic, we report here some recent results from our group. One of the first experiments was carried out at the LOA facility in Palaiseau (France) [16]. In that experiment, highly collimated bunches of high energy electrons were produced by focusing super-intense femtosecond laser pulses at an intensity of $8 \times 10^{19}\,\mathrm{W\,cm^{-2}}$ in sub-millimeter under-dense plasmas. In particular, a thin plastic foil was exploded by the pedestal (lasting several nanoseconds) of the ultrashort pulse, thus producing a pre-plasma suitable for the quasi-resonance of the 35 fs main laser pulse with the electron plasma waves. The density of the plasma was mapped using Nomarski interferometry. The electron beam was fully characterized: up to 10^9 electrons per shot were accelerated, most of which along the laser axis in a beam of aperture below 10^{-3} sterad, with energies up to 40 MeV. Figure 9.1 shows the electron spectrum from this experiment. The measurements were well modeled by 3D numerical simulations, validating a reliable method to generate ultra-short and ultra-collimated electron bunches.

Although very stimulating for the groups active in the field of LPA, experiments performed using thin foils as targets were soon recognized as having a

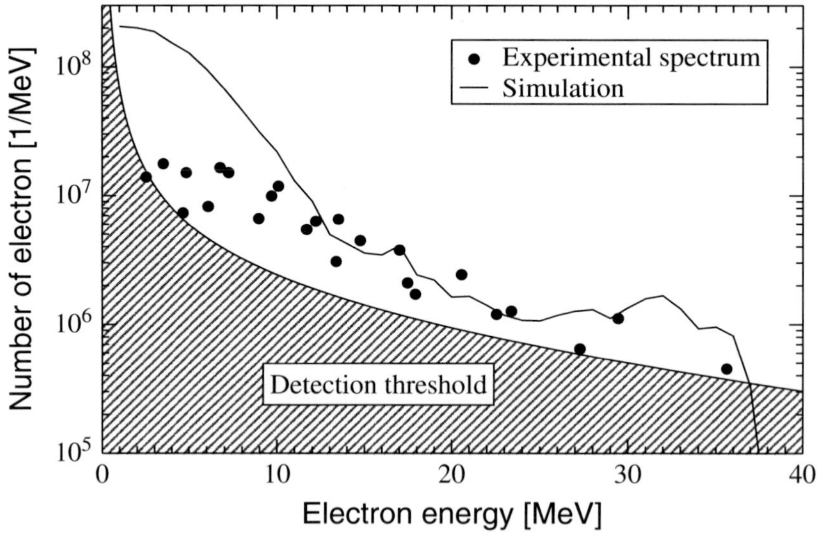

Fig. 9.1. Spectrum of the energetic electrons in the collimated beam as obtained in the experiment described in [16]. *Black dots* represent the experimental data. Also shown is the detection threshold of the electron spectrometer. The *solid line* is the spectrum of the electrons in the collimated beam as obtained from the PIC simulation

major drawback when high average flux particle (electrons or photons) sources have to be built. Indeed, solid targets are not suitable to fully exploit the 10 Hz repetition rate potential of most of the table-top joule class TiSa laser systems available today, as the requirement for a fresh surface at each shot cannot be easily fulfilled. Thus, gas-jet targets are now mostly employed for LP electron acceleration experiments, which can in principle run at a few tens of hertz. In the last few years, a highly nonlinear regime of laser–plasma interaction was studied, which gives rise to a highly electron depleted, nearly spherical, region, called *bubble*, where self-injected electron bunches are subsequently accelerated up to very high energies (without the need for laser guiding by preformed plasma channels) and quasi-monochromatic electron beams (with energy spreads of the order of some %) can in principle be obtained [17–19]. The electron trapping mechanism in this regime is somehow different from the one due to just longitudinal or transverse wavebreaking. As clearly showed in the 3D PIC simulations reported in [17], the trajectories of the electrons, which experience both the ponderomotive force of the laser pulse and the Coulomb field due to the bubble, are such that electrons initially close to the propagation axis accumulates just at the rear of the bubble itself, thus getting trapped and being accelerated. This regime is particularly interesting, as it enables to foresee an actual chance of going in the near future from LP acceleration experiments to real LP based accelerators, with electron bunch parameters, such as energy spread and emittance, suitable for high energy particle experiments [20]. Furthermore, in the presence of an ion channel, the transverse motion of the electrons during their acceleration phase can give rise to the production of ultrashort and ultraintense X-ray beams, as it will be discussed in Sect. 9.4.

From an experimental viewpoint, using simple gas-jet targets, electron beams with energy up to a few hundreds of MeV, accelerated over millimeter distances, have been produced so far, with energy spread of the order of 5–10% and total charge of some hundreds of picocoulombs [7, 21, 22]. In a recent experiment, the use of a capillary to guide the laser pulse over long distances at an intensity suitable for the excitation and the propagation of the plasma wave allowed to get a 1 GeV electron beam, with an acceleration length of 33 mm [23]. In that experiment, the laser pulse was successfully guided through a 400 μm diameter H_2 filled capillary discharge waveguide, made up of alumina. Parameters of the waveguide size, the gas pressure, and the discharge current were previously optimized so as to have a good guiding of the main laser pulse, with very small losses [24].

Recently, a successful experiment [25] using a two pulses scheme has been reported, designed in particular to have a better control of the electron injection into the plasma wave, thus improving the stability and the control of the electron bunch parameters such as energy and energy spread [26, 27]. In detail, two counter-propagating laser pulses were used in the experiment: the main one, that is the one with the greater amplitude, creates a wakefield, while the second one was used for injecting electrons into the wakefield with

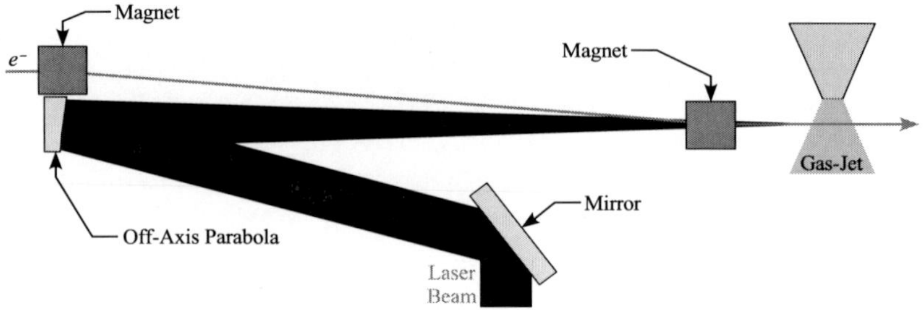

Fig. 9.2. Schematic setup for the external injection of electrons from a LINAC into a laser wakefield

the proper phase and in the proper place in the plasma. Indeed, in the region where the two beams crosses each other, a standing wave is created which, due to the radiation pressure, can inject very small bunches of electrons into the wake of the first pulse. The method, although leading to small total charge in the electron bunch (of some tens up to a few hundreds of pC), allows in principle a fine control of the electron energy and to get very small values of the bunch duration (of the order of 10 fs) and of the energy spread, of the order of 1%, by properly tuning the plasma parameters and the time delay between the two pulses.

However, it should be noted here that the requirements for the Free-Electron-Lasers and high energy accelerators for particle physics experiments are still more stringent. As for the maximum energy, two-stage LP accelerators have been considered so far. However, the most attractive answer to the requirements by high energy physicists seems to be the external injection of relativistic electron bunches from a LINAC into a plasma wave (see Fig. 9.2 for a simple scheme). Although very recently an experiment has been reported using a *plasma wakefield* scheme [28], no experiments have been carried out so far using a Laser Wakefield Acceleration stage to accelerate electrons from a LINAC. Actually, this is a very challenging task: even from a simple argument one can realize that the injection of an electron bunch into plasma waves with wavelength smaller than 100 μm would require a laser and a LINAC synchronized with a jitter of the order of 10 fs! We mention here that recently a project (PLASMONX) was funded by the Instituto Nazionale di Fisica Nucleare (INFN) in Italy devoted to such a kind of experiments: a 300 TW laser system will operate in the near future at the Laboratori Nazionali di Frascati, synchronized with a 150 MeV LINAC.

As an example of an experimental setup needed for the kind of experiments discussed above, we give now a brief overview of a recent experiment carried out by our group at the SLIC facility at CEA in Saclay (France) [10]. The laser system was the UHI-10 Ti:Sa laser, which delivered 65 fs pulses with energy up to 0.7 J. The laser beam was focused by an $f/5$ off-axis parabolic mirror, producing a quasi-gaussian spot where the field parameter $a_0 = eA_{\mathrm{laser}}/m_{\mathrm{e}}c^2$, where A_{laser} is the amplitude of the vector potential of the laser pulse, was $\simeq 2$. We recall here that the value $a_0^2/2$ gives the ratio of the electron quiver energy in the laser electric field to its rest energy. The ns-scale contrast of the pulse $>10^6$ ensured that no pre-plasma was formed by the amplified spontaneous emission (ASE). Supersonic gas jet nozzles of different diameters were tested in the experiment, and an optimum condition in terms of electron bunch total charge was found for 4 mm nozzle with 25 bar backing pressure for H_2 gas. This corresponds to an atomic density of $10^{19}\,\text{cm}^{-3}$ at 0.5 mm from the nozzle exit, that is, at the position of the laser axis.

The electron density of the produced plasma was diagnosed by means of a Mach–Zehnder interferometer, operated with a small portion of the main pulse, doubled in frequency [29,30]. The electron density along the pulse path was measured to be $n_{\text{e}} \approx 2 \times 10^{19}\,\text{cm}^{-3}$. At this density, the electron plasma wave has a period $T_{\text{p}} \approx 25$ fs and wavelength $\lambda_{\text{p}} \approx 7.5\,\mu\text{m}$.

The electron bunches were characterized by using four different diagnostics. A Lanex screen, placed at a distance of 44 mm from the focal plane, provided the overall spatial distribution of the electrons. A magnetic spectrometer, with a Lanex screen as a detector, provided the shot-to-shot electron energy spectrum. A spatial high energy electron beam analyzer (SHEEBA) [31] device provided the angular distribution of each spectral component of the electron bunch. Finally, a photo-nuclear activation setup via induced Bremsstrahlung provided a precise measurement of the electron bunch charge via Monte Carlo code deconvolution. The Lanex screen was essential for tuning the electron accelerator; once tuned, the electron bunch parameters including divergence, spectrum, and charge were measured with SHEEBA, magnetic spectrometer, and nuclear activation. SHEEBA raw data were analyzed with an original algorithm [31] based on the particle transport Monte Carlo libraries GEANT4 [32]. The algorithm transforms the raw patterns produced by the electrons in the radiochromic films (RCF) into the spatial distribution of electrons for each energy component; as an example, Fig. 9.3 shows three data retrieved in this way, at an electron energy of 12, 18, and 25 MeV, respectively. The data analysis of all the RCFs showed an electron energy spectrum peaked at 12 MeV, 8 MeV in width. Furthermore, the electrons showed a divergence of about 30 mrad FWHM, while the pointing direction of the electron beam over several tens of shots was within an angle of less than 100 mrad FWHM.

9.3 Protons Sources

Because of their inertia, the effect on the ions of the high frequency electric field of the electron plasma waves dealt with so far averages out to zero. However, surprisingly, light ions and/or protons have been observed in ultrahigh intensity laser pulse interaction with solid targets since the late 1990s [33–36]. Later on, these particles were found to come from contaminant layers on

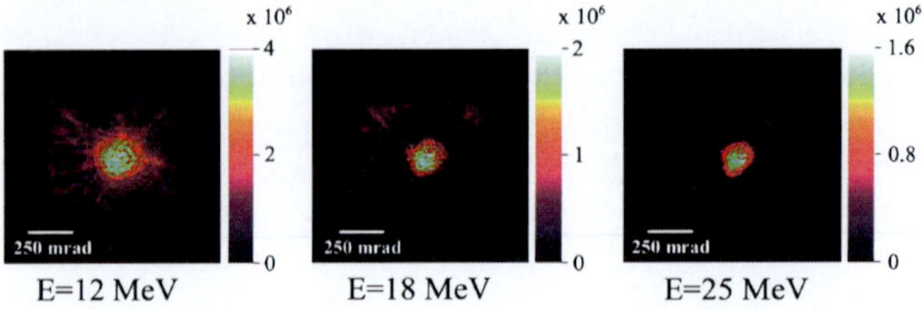

Fig. 9.3. Spatial distribution of the electrons at three different energies as obtained in the experiment described in [10]

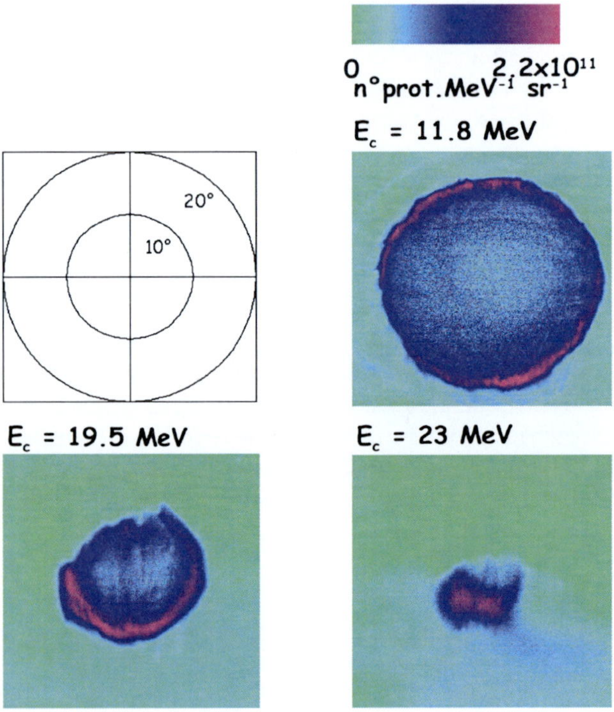

Fig. 9.4. Spatial distribution of the protons obtained in an experiment carried out at the Vulcan facility at Rutherford Appleton Laboratory, retrieved with the diagnostics described in [31]

the target surface. In particular, energetic ions were detected from both the front and the rear side of the target when subpicosecond pulses were used at intensities exceeding $\sim 10^{18}\,\mathrm{W\,cm^{-2}}$. Although different mechanisms have been suggested since then, the so-called target normal sheath acceleration (TNSA) model [8, 37] is currently accepted as the main process accounting for this ion acceleration. Within this mechanism, part of the laser energy is transferred to one or more populations of hot electrons through either ponderomotive or other thermal processes [38]. When propagating across the plasma–vacuum interface, either on the front or on the rear target side, these electrons then give rise to a thin sheath, with a size of the order of the Debye length, where intense electrostatic fields, of the order of 1 TV m^{-1}, can build up, which in turn are able to accelerate the ions. Up to now, ions accelerated up to a few tens of MeV have been reported. In particular, in 2000, proton energy up to 58 MeV have been observed by using the Nova PW laser at LLNL [36], while protons with energies up to a few tens of MeV are currently produced with 100 TW class laser systems. Efforts devoted to increase the proton energy by using ultrathin foils [39–41] or to improve their monochromaticity by suitable target design [42, 43] are now going on [44].

Several diagnostics have been employed to detect and characterize the accelerated ions/protons, among which are Thomson parabolas, CR39 films, and RCFs. In particular, this latter kind of detector can be used in a stacked arrangement, as for the SHEEBA detector discussed above for the electrons. As an example, Fig. 9.4 shows a result from an experiment performed by our group; as it is clearly apparent, more and more energetic protons are emitted in a progressively narrower angle along the laser beam axis.

Because of the intrinsic properties of the proton bunches emitted after ultrahigh intensity laser–solid interaction, such as their laminarity and duration [45], proton bunches have been successfully used for the detection of transient magnetic and electric fields in laser–plasma interaction [46]. In practice, a laser plasma proton bunch can be thought as emitted by a point source before the front side of the target, so that studying the proton deflection allows to retrieve the electric or magnetic field properties in a radiographic fashion. Furthermore, laser plasma protons are also under study for the fast ignition scheme in ICF [47].

Beside to these physical applications, proton beams from LPP are under consideration for their possible use in the medical field; indeed, although the maximum energy is currently too low, properties such as emittance or current can be comparable or even better than for conventional RF accelerators [48]. The most attractive application is probably the hadron therapy [49]. It relies on the specific way by which hadrons deposit their energy in matter: as it is well known, hadrons preferentially release energy at a well defined depth, which depends on their initial energy. The presence of this depth is referred to as the *Bragg peak*. This absorption behavior makes thus possible the ion/proton bombardment of inner tumors, with no or low damages to the surrounding tissues. Roughly speaking, once the hadron beam energy has been fixed, almost all its energy will be released at a given depth where the tumor is located. At the present, figures such as the proton beams energy, energy spread, and shot-to-shot reproducibility makes it unfeasible to use them in actual cancer treatments. However, because of the enormous advantages with respect to conventional accelerators, in terms of both costs and size, several initiatives are actively going on, aimed to make this technique suitable for applications in clinical practice.

9.4 Laser Plasma Based e.m. Sources

In this section a brief survey will be given of the completely new generation of ultrashort and ultrabright LP based e.m. sources, which can be foreseen for the near future as emerging from the latest few years' theoretical, numerical, and experimental studies. Only those issues related to the so-called ultrafast e.m. sources will be concerned here, that is, whose duration can in principle range from some tens up to a few hundreds of femtoseconds. This duration is in principle comparable to the one of the ultrashort driving laser pulses, based on the CPA technique. Actually, as it is well known, laser-produced plasmas have been recognized as very interesting sources, in terms of brightness and short duration, of UV and X-ray radiation since the earliest studies based on nanoseconds pulses. This can be understood as the high electron temperatures and the high ion charge states reached are considered, which lead to a strong e.m. emission arising from electron–ion recombination, ionic transitions, or Bremsstrahlung due to collisions between thermal electrons and ions. Such emission processes usually occur on the timescale of some tens of picoseconds up to a few nanoseconds. A thorough review of these emission processes and their features has been given in [50], so that it will be skipped here.

The process earliest recognized as a potential source of ultrafast X-ray pulses from laser-produced plasmas is the so-called Kα emission [51, 52]. The primary process for this kind of X-ray production from intense laser irradiation of solids is the generation of one or more populations of fast, or hot, electrons having energies of the order of some tens to some hundreds of keV [53–56]. Several mechanisms can be responsible for the generation of such electrons, depending on the initial plasma scalelength and laser pulse intensity, such as resonance absorption, vacuum heating, or ponderomotive acceleration [38, 57–61]. These energetic electrons can then penetrate into the underlying solid target material, where they knock out electrons preferentially from the inner electronic shells of the atoms or ions [62–64]. The radiative transitions of electrons from the outer shells finally lead to the generation of characteristic K lines. In this way, efficient ultrashort X-ray sources at selected photon energies from some keV up to some tens of keV, depending upon the employed target atomic number, can be obtained. Because of this basic production process, the study of the Kα emission is currently adopted as the main diagnostics of the fast electron transport inside the overdense matter. We do not further discuss this issue here, as it is discussed in the work by Gizzi in this volume.

One of the first simulation/theoretical study of the Kα emission dynamics was given by Reich et al. in 2000 [65]. In that work, Kα pulse durations down to $\approx$100 fs were predicted to be achievable, by using thin foil targets of suitable thickness to minimize the fast electrons transit time inside the target.

Concerning the Kα yield, the experimental works published in the last years showed a conversion efficiency of laser energy to Kα photons of 10^{-4}–10^{-5}, leading to a total number of 10^9–10^{10} Kα photons per joule of laser energy for each driving laser shot. As an example, Fig. 9.5 from [66] shows

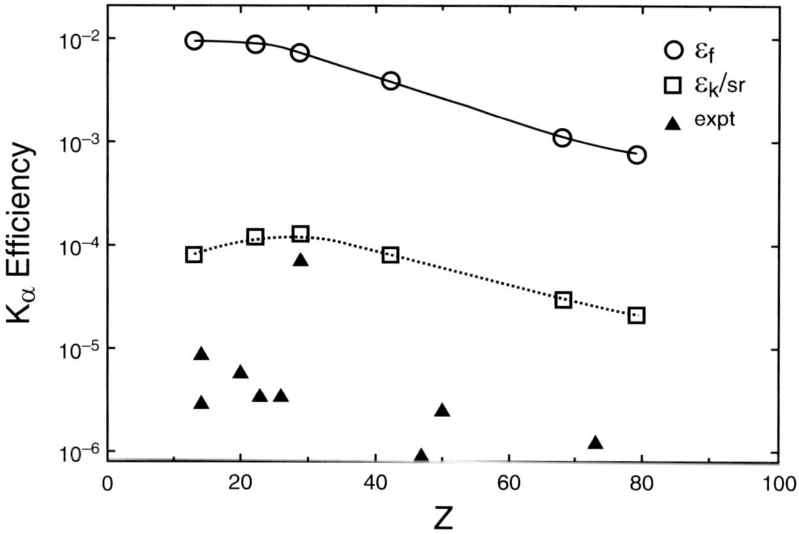

Fig. 9.5. Kα production efficiencies under optimal conditions for various target materials Z, as reported by [66]: photons per electron (*circles*); energy conversion efficiency per srad (*squares*); and experimental measurements (*filled triangles*)

some experimental data points from different experiments (see that paper for details). Similar values have been reported recently, even with laser pulses as short as 45 fs (see [67]). Figure 9.5 also clearly shows the significant overestimation of the Kα conversion efficiency predicted by numerical simulations as compared to the experimental values, at least in those case. Actually, as it came out during the last years [67], the simulation of the whole Kα emission process is quite a complex task, as it requires simultaneously the laser light absorption and the subsequent hot electrons generation and their transport [68] inside the target to be modeled. Issues such as the presence of ASE pedestals or ps prepulses, affecting the plasma scalelength and the laser energy overall absorption, as well as electron transport inhibition have been recognized as crucial in the whole Kα emission process [69]. For instance, a number of different scaling laws for the hot electron temperature as a function of the laser intensity (through the $I\lambda^2$ parameter) have been given, due to the different initial plasma scalelengths and to the different hot electrons production processes involved [65, 70–72]. Furthermore, the role of the hot electron recirculation inside the target and of the hot electrons transport inhibition by the electric fields [73, 74] have to be taken into account for a correct estimate of the total Kα conversion efficiency.

Ultrashort laser–plasma X-ray sources based upon the Kα emission process have been exploited since about 10 years ago to study some dynamical effects in simple systems, typically involving atomic motion, occurring on the timescale of ∼100 fs. A complete review of such effects, in particular in the field of solid state physics and molecular biology, has been given in [75]. Since then, a lot of photoinduced ultrafast processes [76–78] have been studied in the so-called pump and probe class of experiments, which directly exploits the intrinsic synchronization between an optical pump pulse (typically a portion of the laser pulse driving the Kα emission), used to trigger the desired reaction, and an X-ray probe pulse. Ultrashort laser–plasma X-ray sources, based either on Kα or on Bremsstrahlung hard X-ray emission, have also been considered for medical applications in the field of radiography and coronary angiography [79,80]. In this case, besides their short duration and small source size, LP Kα sources have been considered for the simultaneous presence of different fluorescence lines, allowing, by using suitable Bragg or Laue crystals, to detect selected tracers by means of the so-called dual energy subtraction technique. Finally, it should be noted that in the last few years the spatial coherence properties of LPP ultrashort X-ray sources have been investigated, in view of phase-contrast imaging applications [81,82].

LP X-ray sources based on the Kα emission process are the only ultrashort sources that have been actually used for applications right now. However, they suffer from two major drawbacks, which may limit their use in some cases: (1) the emission process is isotropic (hence the figure for the Kα yield given above is for an emission over 4π srad) and (2) the photon energy can only be "tuned," over a discrete set of emission lines by changing the emitter, that is, the target material. The former issue requires bent Bragg (or Laue) crystals to be used in actual experiments, in order to collect as much X-ray radiation as possible; due to the crystal geometry and efficiency issues, this leads to a strong reduction of the number of the available photons on the sample, from 10^9–10^{10} for a joule class ultrashort laser (see above) to about 10^4–10^5.

A completely different concept of ultrashort X-ray source has been discovered and studied in the last few years, based on the so-called betatron emission from relativistic electrons propagating in an ion channel [83, 84]. As it was said in Sect. 9.2, the ponderomotive force of an ultrashort and ultraintense laser pulse expels electrons from the propagation axis, creating a plasma channel and eventually leaving a wake behind the pulse itself. Thus, the electrons trapped in an accelerating period of the wake and injected off-axis will experience, besides an accelerating force, a restoring force directed toward the laser beam axis, due to the space charge separation, and they will thus undergo a transverse oscillation. The restoring force can be quite high in the bubble regime cited in Sect. 9.2. The situation is thus quite similar to the one encountered in a synchrotron radiation facility, where an electron beam is forced to oscillate in the transverse direction by a magnetic field generated by insertion devices such as wigglers or undulators. The role of the magnetic field force is played here by the Coulomb force in the channel. In fact, an analogous of the *wiggler parameter* can be defined in this case, given by $K = \gamma k_b r_0 = 1.33 \times 10^{-10}\sqrt{\gamma n_e[\mathrm{cm}^{-3}]}r_0[\mu\mathrm{m}]$ [84]. Here γ is the electron gamma factor, r_0 is the amplitude of the betatron oscillation, and $k_\mathrm{b} = 2\pi/\lambda_\mathrm{b}$, with λ_b period of a betatron oscillation, given by

$\lambda_b = 2\pi\sqrt{2\gamma}/(c/\omega_p)$. As it is well known from synchrotron theory (see for example [85]), for small ($K < 1$) values of the wiggler parameter, the radiation mainly occurs at one fundamental frequency, while for high values of K ($K \gg 1$), a broadband continuum is emitted. In this latter case, it is useful to define a critical frequency given, in the case of relativistic electrons in a plasma, by $\hbar\omega_c$ [eV] $= 5 \times 10^{-24}\gamma^2 n_e$ [cm^{-3}]r_0 [µm]. The emitted spectrum then increases as $\omega^{2/3}$ for $\omega \ll \omega_c$, reaches a maximum at $0.29\,\omega_c$, and then exponentially decreases to zero above ω_c. The divergence of the emitted radiation is related to the maximum excursion angle of the accelerated electrons during their oscillations, roughly given by $\theta_{max} \simeq K/\gamma$. The total X-ray yield is of course related to the number of betatron oscillations experienced by the electrons: $N_x \simeq 5.6 \times 10^{-3} N_0 K$, where N_0 is the number of betatron oscillation performed by an electron. Finally, the X-ray bunch duration can be expected to be of the order of the electron bunch acceleration duration, thus of the order of 10 fs.

In the last few years, betatron X-ray beams with energy up to a few keV have been observed, which were attributed to electrons accelerated up to some tens of MeV, undergoing betatron oscillations with amplitude $r_0 \simeq 1$ µm [86]. The divergence of a betatron X-ray source is of some tens of mrad. This last figure makes X-ray sources based upon betatron emission very attractive, due to the X-ray flux available for applications. Indeed, 3D PIC simulations show that an X-ray flux of 10^6–10^7 photons per shot, in an energy interval from a few keV up to 100 keV, can be easily obtained with the 100 TW class lasers, which are now starting to be operated worldwide [9]. While smaller than the total X-ray flux available with Kα sources, this figure leads to a higher average brightness, due to the much smaller divergence (Kα emission occurs over 4π sterad). As a result, an average brightness of 10^9–10^{10} ph/s/mm^2/mrad^2/0.1%BW can be foreseen for the betatron sources in the next few years (the corresponding value for the Kα sources is 10^7–10^8).

It is worth noting at this point that ultrashort, high brightness visible beams have been recently obtained, with a table-top equipment based upon a laser–plasma accelerator, by using a small external undulator [87]. In particular, relativistic electron bunches produced by a LWFA scheme, with energy in the 55–75 MeV range, have been forced to undergo transverse oscillations in a 1 m long external undulator, with $K = 0.6$. Starting with electron beams with charge of some tens of pC and an energy spread of about 5%, photon beams in the visible range with bandwitdh of about 50 nm have been obtained in this way, with a brightness of about 10^{16} ph/s/mm^2/mrad^2/0.1%BW. Scaling this values with the figures obtained with the experiment by Leemans et al. cited above [23] would allow photon beams with energy of some hundreds of eV to be obtained, with peak brightness of the order of 10^{24} ph/s/mm^2/mrad^2/0.1%BW.

One of the first proposed processes for obtaining electromagnetic radiation up to the X/γ-ray range using laser beams was the Thomson scattering (see e.g., [88] and references therein). As it is known, this is the scattering of a photon by an electron in which the energy of the final photon is much lower than the electron rest energy. The energy and angular distribution of the incoherent scattered photons depend on the initial photon and electron energy as well as on the photon density. As an example, if a linearly polarized laser beam with normalized amplitude $a_0 \ll 1$ (*linear* regime) and frequency ω_{L} interacts at an angle α with a beam of relativistic electrons with $\gamma_{\mathrm{e}} \gg 1$, the resulting back-scattered radiation has a peak frequency of $\omega_{\mathrm{TS}} \simeq 2\gamma_{\mathrm{e}}^2\omega_{\mathrm{L}}$ $(1 - \cos\alpha)$ and is mainly emitted in the forward direction of the electron beam in a cone with aperture $1/\gamma_{\mathrm{e}}$. Thus, if the laser beam and the electron beam propagate head-on, a beam of photons with frequency peaked at $\omega_{180°} \simeq 4\gamma_{\mathrm{e}}^2\omega_L$ is emitted in a narrow cone in the electron direction. As a_0 increases (so that the so-called *nonlinear* regime is approached), harmonics of the above fundamental frequency appears due to the complex motion of the electrons in the laser field [89] and the angular distribution becomes more and more complex (see e.g., [88]); in the limit $a_0 \gg 1$, a near continuum of high-harmonic radiation is emitted (the parameters a_0 thus plays a role similar to the wiggler parameter K).

Thomson scattering sources are currently considered as very promising X-ray sources, both monochromatic (at least in the linear regime) and easily tunable by changing the electron beam energy. Furthermore, Thomson scattering radiation has been proposed as a tool to diagnose the spectrum and angular distribution of relativistic electron beams [90].

From an experimental viewpoint, Thomson scattering has been recently observed by using electron beams from a LINAC [91] as well as from a laser–plasma source based on LWFA [92]. This latter observation, in particular, allows the availability of table-top, tunable sources of X/γ-rays to be foreseen.

Finally, for the sake of a rough comparison with the corresponding figures for the Kα and betatron sources, we mention here that, as an example, a number of photons up to about 10^9, with mean energy of 20 keV and energy spread of about 10%, can be estimated for the conventional (that is, based on the laser photons scattering off LINAC accelerated electrons) Thomson scattering source in the PLASMONX project cited above [93]. The pulse duration is of some picoseconds, while the source size is estimated to be of about 10 μm and the X-ray beam divergence of some mrad.

9.5 Conclusions

The overwhelming development of the ultrashort and ultraintense laser systems based on the CPA technique enables to foresee a clear way to get miniaturized (table-top) particle accelerators and related sources of X/γ-rays. Such a reduction in dimensions and costs has made realistic the application of these innovative apparatus to several fields, from medical diagnostics and therapy to material science and femto-chemistry. Furthermore, particles accelerated by laser–plasma interaction can be used in the fast-ignition approach to ICF. For this reasons, national (such as the PLASMONX strategic project by the Italian INFN) as well as international efforts (such as the European project ELI) are now in progress to fully investigate the new physics that the laser–matter interaction at such unprecedented intensities will open up.

Acknowledgements

Part of this work was supported by the INFN project PLASMONX. The access to the CEA facility was supported by the EU scheme for the access to large scale facilities within the LASERNET network. One of us (L.L.) acknowledges support from the Italian MIUR project FIRB "BLISS." The authors also acknowledge support from the Italian Ministry of Education and Research through the project FISR (law 449/97) called "Impianti innovativi" and the project "SPARX."

References

1. T. Tajima, J.M. Dawson, Phys. Rev. Lett. **43**, 267–270 (1979)
2. P. Sprangle, E. Esarey, A. Ting, Phys. Rev. A **41**, 4463–4469 (1990)
3. J.M. Dawson, Phys. Rev. **113**, 383–387 (1959)
4. E. Esarey, P. Sprangle, J. Krall, A. Ting, IEEE Trans. Plasma Sci. **24**, 252–288 (1996)
5. G. Bonnaud, H.S. Brandi, C. Manus, G. Mainfray, T. Lehner, Phys. Plasma **1**, 968–989 (1994)
6. E. Esarey, P. Sprangle, J. Krall, A. Ting, IEEE J. Quant. Electron. **33**, 1879–1914 (1997)
7. C.G.R. Geddes, C. Toth, J. van Tilborg, E. Esarey, C.B. Schroeder, D. Bruhwiler, C. Nieter, J. Cary, W.P. Leemans, Nature **431**, 538–541 (2004)
8. S.C. Wilks, A.B. Langdon, T.E. Cowan, M. Roth, M. Singh, S. Hatchett, M.H. Key, D. Pennington, A. MacKinnon, R.A. Snavely, Phys. Plasma **8**, 542–549 (2001)
9. A. Rousse, K. Ta Phouc, R. Shah, R. Fitour, F. Albert, Eur. Phy. J. D **45**, 391–398 (2007)
10. A. Giulietti, N. Bourgeois, T. Ceccotti, S. Davoine, S. Dobosz, P. D'Oliveira, M. Galimberti, J. Galy, A. Gamucci, D. Giulietti, L.A. Gizzi, D.J. Hamilton, E. Lefebvre, L. Labate, J.R. Marquès, P. Monot, H. Popescu, F. Réau, G. Sarri, P. Tomassini, P. Martin, Phys. Rev. Lett. **101**, 105002 (2008)
11. F. Amiranoff, D. Bernard, B. Cros, F. Jacquet, G. Mathieussent, P. Miné, P. Mora, J. Morillo, F. Moulin, A.E. Specka, C. Stenz, Phys. Rev. Lett. **74**, 5220–5223 (1995)
12. C.E. Clayton, K.A. Marsh, A. Dyson, M. Everett, A. Lal, W.P. Leemans, R. Williams, C. Joshi, Phys. Rev. Lett. **70**, 37–40 (1993)
13. P. Sprangle, E. Esarey, J. Krall, G. Joyce, Phys. Rev. Lett. **69**, 2200–2203 (1992)
14. T.M. Antonsen, Jr., P. Mora, Phys. Rev. Lett. **69**, 2204–2207 (1992)
15. A. Modena, Z. Najmudin, A.E. Dangor, C.E. Clayton, K.A. Marsh, C. Joshi, V. Malka, C.B. Darrow, C. Danson, D. Neely, F.N. Walsh, Nature **377**, 606–608 (1995)

16. D. Giulietti, M. Galimberti, A. Giulietti, L.A. Gizzi, R. Numico, P. Tomassini, M. Borghesi, V. Malka, S. Fritzler, M. Pittman, K. Ta Phouc, A. Pukhov, Phys. Plasma **9**, 3655 (2002)
17. A. Pukhov, J. Meyer-ter-Vehn, Appl. Phys. B **74**, 355–361 (2002)
18. I. Kostyukov, A. Pukhov, S. Kiselev, Phys. Plasma **11**, 5256–5264 (2004)
19. A. Pukhov, S. Gordienko, S. Kiselev, I. Kostyukov, Plasma Phys. Contr. Fusion **46**, B179–B186 (2004)
20. W. Lu, M. Twoufras, C. Joshi, F.S. Tsung, W. B. Mori, J. Vieira, R.A. Fonseca, L.O. Silva, Phys. Rev. Special Topics – Accelerators and Beams **10**, 061301 (2007)
21. S.P.D. Mangles, C.D. Murphy, Z. Najmudin, A.G.R. Thomas, J.L. Collier, A.E. Dangor, E.J. Divall, P.S. Foster, J.G. Gallacher, C.J. Hooker, D.A. Jaroszinski, A.J. Langley, W.B. Mori, P.A. Norreys, F.S. Tsung, R. Viskup, B.R. Walton, K. Krushelnick, Nature **431**, 535–538 (2004)
22. J. Faure, Y. Glinec, A. Pukhov, S. Kiselev, S. Gordienko, E. Lefebvre, J.-P. Rousseau, F. Burgy, V. Malka, Nature **431**, 541–544 (2004)
23. W.P. Leemans, B. Nagler, A.J. Gonsalves, C. Toth, K. Nakamura, C.G.R. Geddes, E. Esarey, C.B. Schroeder, S.M. Hooker, Nat. Phys. **2**, 696–699 (2006)
24. A. Butler, D.J. Spence, S.M. Hooker, Phys. Rev. Lett. **89**, 185003 (2002)
25. J. Faure, C. Rechatin, A. Norlin, A. Lifschitz, Y. Glinec, V. Malka, Nature **444**, 737–739 (2006)
26. D. Umstadter, J.K. Kim, E. Dodd, Phys. Rev. Lett. **76**, 2073–2076 (1996)
27. E. Esarey, R.F. Hubbard, W.P. Leemans, A. Ting, P. Sprangle, Phys. Rev. Lett. **79**, 2682–2685 (1997)
28. I. Blumenfeld, C.E. Clayton, F.-J. Decker, M.J. Hogan, C. Huang, R. Ischebeck, R. Iverson, C. Joshi, T. Katsouleas, N. Kirby, W. Lu, K.A. Marsh, W.B. Mori, P. Muggli, E. Oz, R. H. Siemann, D. Walz, M. Zhou, Nature **445**, 741–744 (2007)
29. A. Giulietti, P. Tomassini, M. Galimberti, D. Giulietti, L.A. Gizzi, P. Koester, L. Labate, T. Ceccotti, P. D'Oliveira, T. Auguste, Ph. Monot, P. Martin, Phys. Plasma **13**, 093103 (2006)
30. L.A. Gizzi, M. Galimberti, A. Giulietti, D. Giulietti, P. Koster, L. Labate, P. Tomassini, Ph. Martin, T. Ceccotti, P. D'Oliveira, P. Monot, Phys. Rev. E **74**, 036403 (2006)
31. M. Galimberti, A. Giulietti, D. Giulietti, L.A. Gizzi, Rev. Sci. Instrum. **76**, 053303 (2005)
32. S. Agostinelli, et al., Nucl. Instrum. Meth. Phys. Res. A **506**, 250–303 (2003)
33. K. Krushelnick, E.L. Clark, M. Zepf, J.R. Davies, F.N. Beg, A. Machacek, M.I.K. Santala, M. Tatarakis, I. Watts, P. Norreys, A.E. Dangor, Phys. Plasma **7**, 2055–2061 (2000)
34. E.L. Clark, K. Krushelnick, M. Zepf, F.N. Beg, M. Tatarakis, A. Machacek, M.J.K. Santala, I. Watts, P.A. Norreys, A.E. Dangor, Phys. Rev. Lett. **85**, 1654–1657 (2000)
35. A. Maksimchuk, S. Gu, K. Flippo, D. Umstadter, V.Y. Bychenkov, Phys. Rev. Lett. **84**, 4108–4111 (2000)
36. R.A. Snavely, M.H. Key, S.P. Hatchett, T.E. Cowan, M. Roth, T.W. Phillips, M.A. Stoyer, E.A. Henry, T.C. Sangster, M.S. Singh, S.C. Wilks, A. MacKinnon, A. Offenberger, D.M. Pennington, K. Yasuke, A.B. Langdon, B.F. Lasinski, J. Johnson, M.D. Perry, E.M. Campbell, Phys. Rev. Lett. **85**, 2945–2948 (2000)

37. P. Mora, Phys. Rev. Lett. **90**, 185002 (2003)
38. S.C. Wilks, W.L. Kruer, IEEE J. Quant. Electron. **33**, 1954–1968 (1997)
39. Q.L. Dong, Z.-M. Sheng, M.Y. Yu, J. Zhang, Phys. Rev. E **68**, 026408 (2003)
40. A.J. MacKinnon, Y. Sentoku, P.K. Patel, D.W. Price, S. Hatchett, M.H. Key, C. Andersen, R. Snavely, R.R. Freeman, Phys. Rev. Lett. **88**, 215006 (2002)
41. T. Ceccotti, A. Lèvy, H. Popescu, F. Rèau, P. D'Oliveira, P. Monot, J.-P. Geindre, E. Lefebvre, P. Martin, Phys. Rev. Lett. **99**, 185002 (2007)
42. B.M. Hegelich, B.J. Albright, J. Cobble, K. Flippo, S. Letzring, M. Paffett, H. Ruhl, J. Schreiber, R.K. Schulze, J.C. Fernàndez, Nature **439**, 441–444 (2006)
43. H. Schwoerer, S. Pfotenhauer, O. Jäckel, K.-U. Amthor, B. Liesfeld, W. Ziegler, R. Sauerbrey, K.W.D. Ledingham, T. Esirkepov, Nature **439**, 445–448 (2006)
44. J. Fuchs, P. Antici, E. D'Humières, E. Lefebvre, M. Borghesi, E. Brambrink, C.A. Cecchetti, M. Kaluza, V. Malka, M. Manclossi, S. Meyroneinc, P. Mora, J. Schreiber, T. Toncian, H. Pèpin, P. Audebert, Nat. Phys. **2**, 48–54 (2006)
45. L. Romagnani, J. Fuchs, M. Borghesi, P. Antici, P. Audebert, F. Ceccherini, T. Cowan, T. Grismayer, S. Kar, A. Macchi, P. Mora, G. Pretzler, A. Schiavi, T. Toncian, O. Willi, Phys. Rev. Lett. **95**, 195001 (2005)
46. M. Borghesi, D.H. Campbell, A. Schiavi, M.G. Haines, O. Willi, A.J. MacKinnon, P. Patel, L.A. Gizzi, M. Galimberti, R.J. Clarke, F. Pegoraro, H. Ruhl, S. Bulanov, Phys. Plasma **9**, 2214–2220 (2002)
47. M. Roth, T.E. Cowan, M.H. Key, S.P. Hatchett, C. Brown, W. Fountain, J. Johnson, D.M. Pennington, R.A. Snavely, S.C. Wilks, K. Yasuike, H. Ruhl, F. Pegoraro, S.V. Bulanov, E.M. Campbell, M.D. Perry, H. Powell, Phys. Rev. Lett. **86**, 436–439 (2001)
48. T.E. Cowan, J. Fuchs, H. Ruhl, A. Kemp, P. Audebert, M. Roth, R. Stephens, I. Barton, A. Blazevic, E. Brambrink, J. Cobble, J. Fernàndez, J.-C. Gauthier, M. Geissel, M. Hegelich, J. Kaae, S. Karsch, G.P. Le Sage, S. Letzring, M. Manclossi, S. Meyroneinc, A. Newkirk, H. Pèpin, N. Renard-LeGalloudec, Phys. Rev. Lett. **92**, 204801 (2004)
49. S. Bulanov, T. Zh. Esirkepov, V.S. Khoroshkov, A.V. Kuznetsov, F. Pegoraro, Phys. Lett. A **299**, 240–247 (2002)
50. D. Giulietti, L.A. Gizzi, La Rivista del Nuovo Cimento **21**, 1–93 (1998)
51. A. Rousse, P. Audebert, J.P. Geindre, F. Falliès, J.C. Gauthier, A. Mysyrowicz, G. Grillon, A. Antonetti, Phys. Rev. E **50**, 2200–2207 (1994)
52. S. Bastiani, A. Rousse, J.P. Geindre, P. Audebert, C. Quoix, G. Hamoniaux, A. Antonetti, J.C. Gauthier, Phys. Rev. E **56**, 7179–7185 (1997)
53. T. Schlegel, S. Bastiani, L. Grémillet, J.-P. Geindre, P. Audebert, J.-C. Gauthier, E. Lefebvre, G. Bonnaud, J. Delettrez, Phys. Rev. E **60**, 2209–2217 (1999)
54. K.B. Wharton, S.P. Hatchett, S.C. Wilks, M.H. Key, J.D. Moody, V. Yanovsky, A. Offenberger, B.A. Hammel, M.D. Perry, C. Joshi, Phys. Rev. Lett. **81**, 822–825 (1998)
55. S. Bastiani, P. Audebert, J.P. Geindre, T. Schlegel, J.C. Gauthier, C. Quoix, G. Hamoniaux, G. Grillon, A. Antonetti, Phys. Rev. E **60**, 3439–3442 (1999)
56. G. Malka, J. Fuchs, F. Amiranoff, S.D. Baton, R. Gaillard, J.L. Miquel, H. Pépin, C. Rousseaux, G. Bonnaud, M. Busquet, L. Lours, Phys. Rev. Lett. **79**, 2053–2056 (1997)
57. P. Gibbon, E. Förster, Plasma Phys. Contr. Fusion **38**, 769–793 (1996)
58. P. Gibbon, A.R. Bell, Phys. Rev. Lett. **68**, 1535–1538 (1992)

59. L.A. Gizzi, A. Giulietti, D. Giulietti, P. Audebert, S. Bastiani, J.-P. Geindre, A. Mysyrowicz, Phys. Rev. Lett. **76**, 2278–2281 (1996)
60. J. Zhang, Y.T. Li, Z.M. Sheng, Z.Y. Wei, X. Dong, Q.L. Lu, Appl. Phys. B **80**, 957–971 (2005)
61. L. Labate, M. Galimberti, A. Giulietti, D. Giulietti, P. Köster, P. Tomassini, L.A. Gizzi, Appl. Phys. B **86**, 229–233 (2007)
62. F. Ewald, H. Schwoerer, R. Sauerbrey, Europhys. Lett. **60**, 710–716 (2002)
63. S.B. Hansen, A.Y. Faenov, T.A. Pikuz, K.B. Fournier, R. Shepherd, H. Chen, K. Widmann, S.C. Wilks, Y. Ping, H.K. Chung, A. Niles, J.R. Hunter, G. Dyer, T. Ditmire, Phys. Rev. E **72**, 036408 (2005)
64. T. Kawamura, H. Nishimura, F. Koike, Y. Ochi, R. Matsui, W.Y. Miao, S. Okihara, S. Sakabe, I. Uschmann, E. Förster, K. Mima, Phys. Rev. E **66**, 016402 (2002)
65. C. Reich, P. Gibbon, I. Uschmann, E. Förster, Phys. Rev. Lett. **84**, 4846–4849 (2000)
66. D. Salzmann, C. Reich, I. Uschmann, E. Förster, P. Gibbon, Phys. Rev. E **65**, 036402 (2002)
67. D. Riley, J.J. Angulo-Gareta, F.J. Khattak, M.J. Lamb, P.S. Foster, E.J. Divall, C.J. Hooker, A.J. Langley, R.J. Clarke, D. Neely, Phys. Rev. E **71**, 016406 (2005)
68. V.T. Tikhonchuk, Phys. Plasma **9**, 1416–1421 (2002)
69. C. Ziener, I. Uschmann, G. Stobrawa, Ch. Reich, P. Gibbon, T. Feurer, A. Morak, S. Düsterer, H. Schwoerer, E. Förster, R. Sauerbrey, Phys. Rev. E **65**, 066401 (2002)
70. H. Nishimura, T. Kawamura, R. Matsui, Y. Ochi, S. Okihara, S. Sakabe, F. Koike, T. Johzaki, H. Nagatomo, K. Mima, I. Uschmann, E. Förster, J. Quant. Spectros. Radiat. Transf. **81**, 327–337 (2003)
71. S.C. Wilks, W.L. Kruer, M. Tabak, A.B. Langdon, Phys. Rev. Lett. **69**, 1383–1386 (1992)
72. G. Malka, J.L. Miquel, Phys. Rev. Lett. **77**, 75–78 (1996)
73. A.R. Bell, J.R. Davies, S. Guerin, H. Ruhl, Plasma Phys. Contr. Fusion **39**, 653–659 (1997)
74. J.R. Davies, Phys. Rev. E **69**, 065402 (2004)
75. A. Rousse, C. Rischel, J.-C. Gauthier, Rev. Mod. Phys. **73**, 17–31 (2001)
76. C. Rischel, A. Rousse, I. Uschmann, P.-A. Albouy, J.-P. Geindre, P. Audebert, J.-C. Gauthier, E. Förster, J.-L. Martin, A. Antonetti, Nature **390**, 490–492 (1997)
77. A. Rousse, C. Rischel, S. Fourmaux, I. Uschmann, S. Sebban, G. Grillon, P. Balcou, E. Förster, J.P. Geindre, P. Audebert, J.C. Gauthier, D. Hulin, Nature **410**, 65–68 (2001)
78. K. Sokolowski-Tinten, C. Blome, J. Blums, A. Cavalleri, C. Dietrich, A. Tarasevitch, I. Uschmann, E. Förster, M. Kammler, M. Horn-von-Hoegen, D. von der Linde, Nature **422**, 287–289 (2003)
79. E. Andersson, G. Hölzer, E. Förster, M. Grätz, L. Kiernan, A. Sjögren, S. Svanberg, J. Appl. Phys. **90**, 3048–3056 (2001)
80. J.C. Kieffer, A. Krol, Z. Jiang, C.C. Chamberlain, E. Scalzetti, Z. Ichalalene, Appl. Phys. B **74**, S75–S81 (2002)
81. R. Toth, S. Fourmaux, T. Ozaki, M. Servol, J.C. Kieffer, R.E. Kincaid, Jr., A. Krol, Phys. Plasma **14**, 053506 (2007)
82. D. Boschetto, G. Mourou, A. Rousse, A. Mordovanakis, B. Hou, J. Nees, D. Kumah, R. Clarke, Appl. Phys. Lett. **90**, 011106 (2007)

83. A. Rousse, K. Ta Phouc, R. Shah, A. Pukhov, E. Lefebvre, V. Malka, S. Kiselev, F. Burgy, J.-P. Rousseau, D. Umstadter, D. Hulin, Phys. Rev. Lett. **93**, 135005 (2004)
84. I. Kostyukov, S. Kiselev, A. Pukhov, Phys. Plasma **10**, 4818–4828 (2003)
85. D. Attwood, *Soft X-rays and Extreme Ultraviolet Radiation* (Cambridge University Press, Cambridge, 1999)
86. K. Ta Phouc, F. Burgy, J.-P. Rousseau, V. Malka, A. Rousse, R. Shah, D. Umstadter, A. Pukhov, S. Kiselev, Phys. Plasma **12**, 023101 (2005)
87. H.-P. Schlenvoigt, K. Haupt, A. Debus, F. Budde, O. Jäckel, S. Pfotenhauer, H. Schwoerer, E. Rohwer, J.G. Gallacher, E. Brunetti, R.P. Shanks, S.M. Wiggins, D.A. Jaroszinski, Nat. Phys. **4**, 130–133 (2008)
88. E. Esarey, S.K. Ride, P. Sprangle, Phys. Rev. E **48**, 3003–3021 (1993)
89. Y.Y. Lau, F. Hei, D.P. Umstadter, R. Kowalczyk, Phys. Plasma **10**, 2155–2162 (2003)
90. P. Tomassini, M. Galimberti, A. Giulietti, D. Giulietti, L.A. Gizzi, L. Labate, Phys. Plasma **10**, 917–920 (2003)
91. M. Uesaka, H. Kotaki, K. Nakajima, H. Harano, K. Kinoshita, T. Watanabe, T. Ueda, K. Yoshii, M. Kando, H. Dewa, S. Kondo, F. Sakai, Nucl. Instrum. Meth. Phys. Res. A **455**, 90–98 (2000)
92. H. Schwoerer, B. Liesfeld, H.-P. Schlenvoigt, K.-U. Amthor, R. Sauerbrey, Phys. Rev. Lett. **96**, 014802 (2006)
93. P. Tomassini, A. Bacci, J. Cary, M. Ferrario, A. Giulietti, D. Giulietti, L. A. Gizzi, L. Labate, L. Serafini, V. Petrillo, C. Vaccarezza, IEEE Trans. Plasma Sci. **36**, 1782–1789 (2008)

10

Laser-Driven Ion Generation with Short, Intense, and High Contrast Pulses

Tiberio Ceccotti, Anna Lévy, and Philippe Martin

Summary. About 10 years ago, first experimental works on high intensity ($>10^{18}$ W cm^{-2}) laser-driven proton acceleration showed the great potential of these ions bunches for different application domains (as, for instance, high resolution radiography, isochore heating, and fast ignition). Moreover, the prospect of a future exploitation of such high energy ions in hadron-therapy certainly further incited the scientific research in this domain. One of the goal of the present research lies in increasing the peak proton energy as this is a critical point for a number of applications. Reducing the target thickness is one of the suggested ways to get higher peak proton energies. Nevertheless, that can be realized using only very high contrast laser pulses.

In this framework, after a brief review about the influence of a plasma gradient on ion acceleration, we will show how we have improved of a factor 10^4 the contrast of our 10 TW laser beam using a Double Plasma Mirror. Then, the results we have obtained in this Ultra High Contrast (UHC) regime concerning the influence of target thickness as well as beam polarization on proton and ion acceleration will be presented. In particular, supported by numerical simulations, we will show that ion bunches with quasi-symmetrical features are emitted from both sides of the target and that their energy is a function of the laser electromagnetic field p component only. Finally, we will present the analytical model we have specially developed to describe ion acceleration properties in the Ultra High Contrast regime.

10.1 Introduction

Undoubtedly, the acceleration of charged particles has been one of the most active research fields in the physics of laser–matter interactions throughout the last 10 years. The phenomenon of laser-driven ion acceleration was well known even before this [1, 2], although essentially limited to the thermal expansion of the coronal plasma, typical of nano and sub-nanosecond low-intensity laser interaction regimes. The technique of CPA, introduced in the late 1980s [3, 4], has enabled the use of shorter pulses and increasingly intense laser beams. Since then, intensities $\geq 10^{18}$W cm^{-2} have been available, allowing the exploration of laser–matter interactions in the so-called "relativistic domain," as the motion of electrons in such an electromagnetic field is relativistic. The huge electric fields produced in plasmas due to the laser-induced charge separation open new fields of research in the domain of charged particle acceleration. Confirming that, high-quality quasi-mono-energetic electron bunches in the hundreds of mega electron volt range have been obtained using one [5–8] or more [9, 10] relativistic laser pulses. With regard to positive charge acceleration, high energy ion (mainly protons) beams have also been produced. The excellent emission properties of these ion beams make them particularly suitable for a wide number of applications, including high resolution probing of electric fields in plasmas [11], fast ignition applications [12], induction of nuclear phenomena [13], isotope production for medical applications [14], and proton therapy [15].

Recent review articles ([16, 17] and references therein) allow the interested reader to get a broader picture of this exciting research domain and related applications. In the following (Sect. 10.2), we will mainly devote ourselves to the principles of the main ion acceleration mechanism, and to the way the temporal profile of the laser pulse, and more specifically the beam contrast ratio, can influence it. In particular, we will briefly review the main theoretical and experimental published work concerning the action of a plasma gradient on ion acceleration characteristics. Section 10.3 presents the contrast improvement device we have implemented for our laser beam, and the related temporal profile measurements. In Sect. 10.4, we will show and discuss the main results obtained using ultra high contrast (UHC) laser pulses in laser-driven ion acceleration experiments. Finally, an example of the exploitation of the particular features of UHC pulses in laser-driven ion acceleration will be given in Sect. 10.5.

10.2 The TNSA Acceleration Mechanism: The Role of a Plasma Gradient

Although the subject of some debate in the early 2000s, the mechanism known as Target Normal Sheath Acceleration (TNSA) is currently recognized by the community as the main acceleration process for the vast majority of presently available laser systems, and has been widely studied experimentally and numerically. The TNSA model was presented for the first time in 2001 by Wilks and co-authors [18], following the first direct observations of intense high-energy proton bunches in 2000 [19–21], although the basic physical principle had already been suggested in 1966 by Gurevich et al. [22]. The TNSA model can be described as follows. We begin by considering a laser beam, with intensity I_L such that $10^{18} \leq I_L \leq 10^{20}\,\mathrm{W\,cm^{-2}}$ and duration $\tau_L \lessapprox ps$, impinging on a solid target of micrometer thickness. For such high intensities, the velocity of the electrons subject to the laser electromagnetic fields becomes comparable to the speed of light. This interaction regime is then called relativistic and corresponds to the condition $a \geq 1$, where the dimensionless parameter a is defined as

$$a = eE_{\mathrm{L}}/m_{\mathrm{e}}c\omega_{\mathrm{L}} \cong 0.85(\lambda_{\mathrm{L}}/\mu m)(I_{\mathrm{L}}/10^{18}W/\mathrm{cm}^2)^{1/2} \tag{10.1}$$

Here $E_{\mathrm{L}}, \omega_{\mathrm{L}}$, and λ_{L} are, respectively, the laser electric field, frequency, and wavelength, while m_{e} and e are the mass and charge of the electron. At such intensities, the laser light can be efficiently absorbed by the target electrons, resulting in a population of hot electrons partly escaping from the irradiated side of the target and partly producing a current streaming into the target. The former set up an electric field accelerating ions in the backward direction (BWD). The electrons propagating into the target first create a space charge field at the critical density surface, pushing front side ions inward. Nonetheless, this ion population (as for that in the BWD) is usually less energetic, less laminar, and more divergent than that produced by the same electrons on the nonirradiated "rear" target surface [23]. Hot electrons having a high enough kinetic energy (and consequently a large enough mean free path) can migrate through the target and exit into vacuum. A kind of electron cloud then forms, confined to the rear side vacuum–target interface by a charge separation field over a distance of the order of the Debye length ($\lambda_{\mathrm{De}} = \sqrt{\varepsilon_0 k_{\mathrm{B}} T_{\mathrm{e_{hot}}}/n_{\mathrm{e_{hot}}} e^2}$, where ε_0 is the electric permittivity and $T_{\mathrm{e_{hot}}}$ and $n_{\mathrm{e_{hot}}}$ are, respectively, the hot electron temperature and density). As a consequence, the strength of the electric field produced in this way will be given by

$$E = \frac{k_{\mathrm{B}} T_{\mathrm{e_{hot}}}}{e\lambda_{\mathrm{De}}}. \tag{10.2}$$

For typical interaction parameters ($\lambda_{\mathrm{De}} \approx 1\mu\mathrm{m}$, $T_{\mathrm{e_{hot}}} \approx$ some hundreds keV), this strong teravolt-per-meter field ionizes atoms in the vicinity of the surface and accelerates them to energies that can be several times larger than the typical electron kinetic energy. Protons contained in hydrocarbons and water vapor impurity layers commonly present on target surfaces are more likely to be accelerated, due to their more favorable mass to charge ratio. As a consequence, a remarkably well collimated, neutralized proton beam is observed in the forward direction, emitted perpendicularly to the rear target surface (FWD). The beam has a short duration ($\approx$ ps), a small emittance, but a large energy spread. The typical efficiency (laser energy to ion energy) is a few percent. The TNSA model refers to this third accelerated ion population.

A large part of experimental and theoretical efforts have been devoted to increasing the maximum attainable proton energy, as this is a priority parameter for a number of applications (the other major trend in present research devoted to proton beam optimization consists of obtaining a "quasi-monochromatic" energy spectral distribution). The 58 MeV proton beam obtained in 2000 at the Lawrence Livermore National Laboratory [19] still stands as the maximum energy ever attained. Actually, this value is still far from, for instance, the approximately 150 MeV required for probing dense core plasmas in Inertial Confinement Fusion Applications [24, 25] or the (at least)

200 MeV required in proton-therapy for an effective treatment of deep seated tumors [26]. Fuchs and co-authors have proposed a scaling law [27], allowing the necessary laser parameters to produce proton beams of interest for such applications to be estimated. In their work, best suited to hundreds of fs/some ps duration laser pulses, they use the self-similar fluid model proposed by Mora [28] giving the following estimate for the maximum FWD proton energy:

$$E_{\mathrm{max}} \sim 2k_{\mathrm{B}}T_{\mathrm{e_{hot}}}\left[\ln\left(t_{\mathrm{p}} + \sqrt{t_{\mathrm{p}}^2+1}\right)\right]^2, \tag{10.3}$$

where $t_{\mathrm{p}} = \omega_{\mathrm{pi}}t_{\mathrm{eff}}/\sqrt{2e^1}$ is the effective acceleration time t_{eff} normalized to the ion plasma frequency ω_{pi} which, considering protons, is given by $\omega_{\mathrm{pi}} = \sqrt{e^2 n_{\mathrm{e_{hot}}}/\varepsilon_0}$. Obviously, expression (10.3) needs a proper choice of the value of t_{eff} to be valuable. Fuchs founded a good agreement with published data by setting $t_{\mathrm{eff}} \sim 1.3\tau_{\mathrm{L}}$ and extrapolating the laser parameters allow us to obtain, for instance, the 200 MeV protons needed for proton therapy ($\tau_{\mathrm{L}} = 0.5$ ps, $I_{\mathrm{L}} = 8 \times 10^{20}$ W cm^{-2}, 3 µm FWHM focal spot with a 10 µm thick target). These values have been revised by Robson [29] who describes the proton acceleration by adding an adiabatic phase of hot electron temperature decrease [30] to the isothermal free expansion phase considered by Fuchs and tries to take into account three-dimensional effects. For the same 200 MeV goal, he finds more strict laser conditions: $I_{\mathrm{L}} = 4 \times 10^{21}$ W cm^{-2} and $\tau_{\mathrm{L}} = 1$ ps for a 25 µm target thickness.

It is worth noting that at such high intensities, acceleration mechanisms other than TNSA may, however, come into play. For instance, using circularly polarized and high contrast beams, it is possible to ponderomotively drive protons (and target ions) from the irradiated target surface (*radiation pressure acceleration*) showing high energies and a narrow spectral distribution [31–33].

Apart from boosting laser performances, another path followed by the community to increase the peak proton energy consists of reducing the target thickness. It is expected [34] that the increase in peak proton energy is inversely proportional to the target thickness. Essentially, the simple 1D model presented in [34] suggests that if the "length" $L_{\mathrm{P}} \sim c\tau_{\mathrm{L}}$ of the hot electron beam flowing through the target is higher than twice the target thickness L, then $n_{\mathrm{e_{hot}}}$ is increased due to electron recirculation. The higher electron density consequently increases the accelerating electric field (Fig. 10.1).

Accordingly, E_{max} is then expected to grow as $L_{\mathrm{P}}/2L$ for $L > (L_{\mathrm{P}}/2)$. Indeed, Mackinnon et al.[35] showed an increase in peak proton energy from 6.5 to 24 MeV by reducing aluminum target thickness from 100 to 3 µm (Fig. 10.2I). Nonetheless, when Spencer et al.[36] investigated the influence of target thickness on peak proton energy using similar conditions (but about one order lower laser intensity), they reached an "optimal" thickness L_{opt} at about 12 µm, with a decrease of E_{max} for thinner targets (see Fig. 10.2II). The presence of an optimal thickness maximizing the peak proton energy and the differences between the two sets of data can be explained in the following way. On a typical CPA laser chain, the main, high-intensity pulse is always preceded

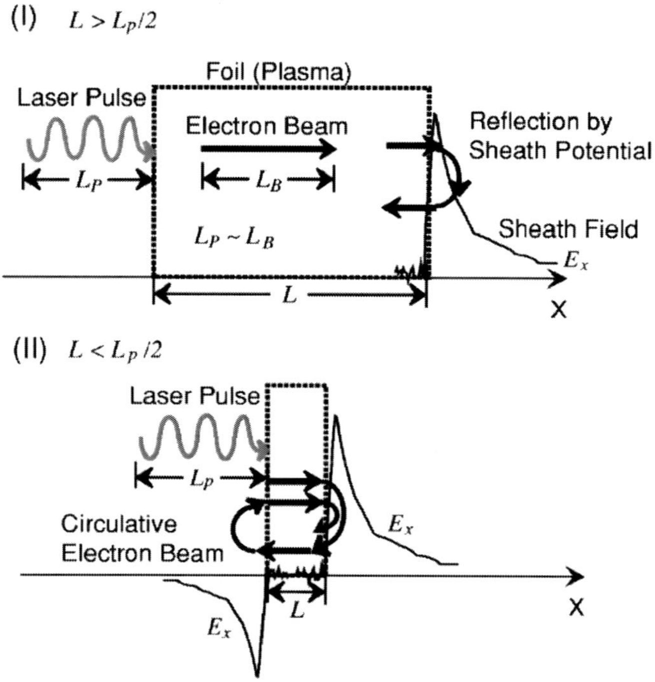

Fig. 10.1. Illustrating $n_{e_{hot}}$ and consequent E field increase due to hot electron recirculation for a thin target (II) compared to a thicker one (I) (Picture extracted from [34])

by a low intensity background made up of pre-pulses (generally due to imperfect matching of some optical element inside the laser chain) and a temporal pedestal. This latter is due to parasitic light known as amplified spontaneous emission (ASE) occurring on a nanosecond time scale and spectral clipping or aberrations of the spectral phase, producing temporal structures on a sub-100 ps time scale. The strength of the background impinging on the target before the arrival of the main pulse can be defined through the "contrast ratio" value. This is defined as the peak intensity of the main pulse divided by the background intensity, and is typically in the 10^5–10^7 range. For a typical peak intensity of $10^{19}\,\mathrm{W\,cm^{-2}}$, pre-pulses and pedestal energy level are then more than enough to heat and ablate the irradiated target surface and form a pre-plasma before the arrival of the main peak. This pre-plasma has two main consequences.

The first is a modification of the laser absorption fraction and hot electron temperature, which turns out to be beneficial to ion acceleration by increasing laser–target coupling. The target conditions affect the transfer mechanism of laser energy into hot electrons. In the case of a steep density gradient, typical of high contrast interactions, vacuum heating [37–39] is expected to be the

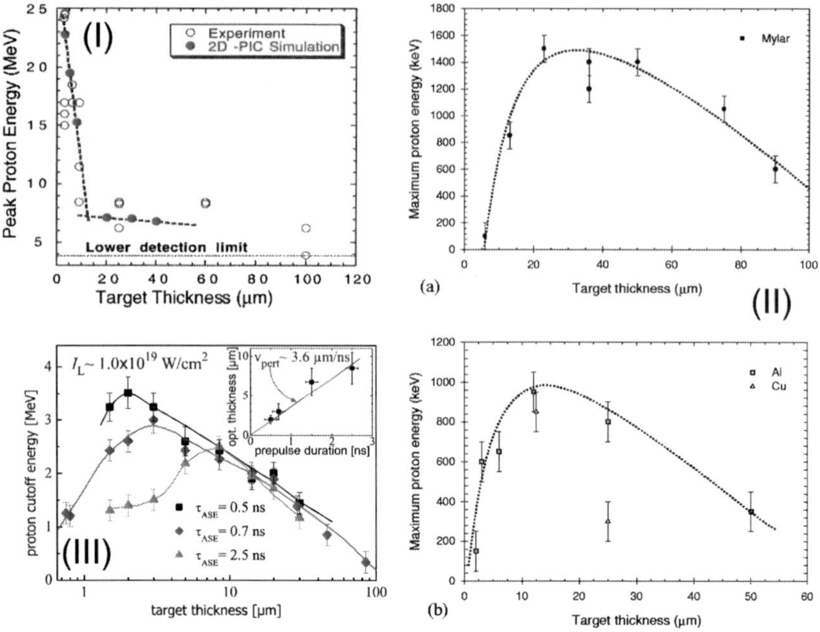

Fig. 10.2. Peak proton energies as a function of target thickness from three different works (see main text). Pictures are extracted from (I) [35], (II) [36], and (III) [43]

main coupling mechanism (or resonant absorption [40] for $a < 1$) (see Sect. 4). For longer gradient scale lengths, several different processes may accelerate electrons, but for usual laser conditions ponderomotive $j \times B$ type mechanisms become the main coupling source of energy transfer [14, 41, 42].

The second consequence is that a shock wave can be launched in the target, with very detrimental consequences for ion acceleration if this shock wave reaches the rear target surface before the main laser pulse hits the front surface. If this happens, the rear surface properties (essentially the presence of a steep density gradient) can be modified by the emerging shock wave, thereby quenching ion acceleration. Moreover, X-ray radiation or thermal conductivity from the pre-plasma can produce plasma rarefaction wave on the nonirradiated face. The maximum proton energy as well as the total extracted charge can be reduced if a density gradient with a scale length $\ell_g > \lambda_{De}$ already exists before the formation of the accelerating plasma sheath. In that case, expression (10.2) must be replaced by

$$E = \frac{k_B T_{e_{hot}}}{e \ell_g} \tag{10.4}$$

As a consequence, the optimal thickness L_{opt} introduced above corresponds, for a given set of laser parameters (pulse duration and intensity, contrast ratio, and pre-pulse duration), to the thickness, allowing the perturbations due to the pre-pulse from the front surface, which travel at typical acoustic speeds of a few µm per ns, to reach the rear target surface only after the hot electron current generated by the main pulse at the front surface. The difference in laser contrast, 10^6 for Spencer [36] and 10^{10} for Mackinnon [35], then explains why the latter was able to shoot thinner targets and achieve higher peak proton energies. Moreover, the difference in L_{opt} between mylar (≈25–30 µm) and aluminum (≈10–15 µm) targets reported in [36] can be explained taking into account the fact that the shock wave propagation speed is lower in aluminum than in mylar (see Fig. 10.2II).

Until now, most published work has been devoted to the influence of a plasma gradient on the irradiated side of the target on the ion emission properties. Kaluza [43] experimentally studied the influence of pre-pulse duration for different target thicknesses, finding that short pre-plasma durations increase peak proton energy and correspond to a reduction of L_{opt} (Fig. 10.2III). His results have been successfully reproduced [44] using an hybrid PIC code to simulate fast electron transport through the pre-plasma. These simulations have also shown a reduction of fast electron collimation propagation with long scale length pre-plasma, and a consequent reduction of the strength of the accelerating electric field on the rear side of the target. Other numerical simulations carried out with a 1D PIC code [45] confirmed that the presence of a pre-plasma can increase proton energy on condition that the maximum density of pre-plasma is close to the critical density, and its scale length is about 60% of the laser pulse FWHM. Long scale length pre-plasma may also influence the ion acceleration direction [46], increasing the influence of hole boring [47] and deviating the ion bunch from the target normal towards the laser direction. Moreover, 2D PIC simulations have shown [48] that for a pre-plasma with a gradient of around $10 \times L_{\mathrm{P}}$, the ion beam collimation is improved and the peak energy of the FWD and BWD beams, created at the front side of the target at the critical density surface ($n_{\mathrm{c}} = \varepsilon_0 m_{\mathrm{e}} \omega_0^2 / e^2$, ω_0 being the pulse frequency), is reduced.

As mentioned earlier, a plasma gradient on the target rear side can affect the ion emission properties. Mackinnon [49] induced a 100 µm scale length plasma on the back side of a 25 µm Al foil with a laser beam 250 ps before the arrival of the main interaction beam. The measurements showed a reduction in peak proton energy of more than 75%. A similar experiment was more recently reported by Fuchs [50]. In that case, a delay line between the pre-forming and the main beam allowed the production of different sized plasmas on the target rear side. The obtained scale lengths had been estimated using a hydrodynamic code in addition to a 1D PIC code, and their values ranged from about 2 to 20 µm. Fuchs successfully reproduced experimental data using a 1D PIC and an adiabatic hybrid code and did not observe any relevant proton energy decrease for scale lengths lower than 5 µm. On the contrary, a small increase in proton energy compared to the case without a gradient was observed for $\ell_{\mathrm{g}} \sim 2$ µm and was attributed to ion wave breaking as described in [51]. Finally, Andreev [52] determined the optimal pre-plasma scale lengths on the front and on the back of the target through numerical simulations and an analytical model. The ideal pre-plasma should create a smooth plasma gradient on the front of the target to maximize the laser energy coupling, while still keeping a sufficiently steep gradient on the target rear side (respectively, about $\sim 4\lambda_{\mathrm{De}}$ and $\sim 2\lambda_{\mathrm{De}}$ for a pure hydrogen plasma with a maximum density $\sim 2n_{\mathrm{c}}$).

This not-exhaustive survey of theoretical and experimental work shows the importance of a deep understanding of the influence of a pre-plasma on laser-driven ion acceleration to improve the qualities of the produced ion beam. This is significant, as the pedestal, at the origin of the pre-plasma, is an intrinsic drawback of usual CPA laser chains and shows very large shot-to-shot intensity variations. This obviously affects the pre-plasma as well the final ion bunch characteristics. These variations add a further difficulty to the interpretation of experimental data using numerical codes or analytical models. It ensues from the above that using "pedestal free" laser pulses has several advantages. For instance, high contrast beams are much less affected by shot-to-shot intensity variations, allowing the generation of highly reproducible ion beams (see the right-hand side of Fig. 10.9 below). That simplifies the related data analysis and allows a better understanding of the main mechanisms of the interaction. Moreover, as seen earlier, removing the pedestal from laser pulses allows the shooting of thinner foils and consequently achieving higher peak proton energies. Because of that, about 4 years ago we started an ambitious upgrade program devoted to improving the contrast of our 10 TW laser beam (about 10^6) by a factor 10^4. In the following, we report how we succeeded in that using a Double Plasma Mirror (DPM) [53] and the main results we have obtained for laser-driven ion acceleration in this Ultra High Contrast interaction regime [54].

10.3 Ultra High Contrast Pulses with a Double Plasma Mirror

Many different ways to remove the temporal pedestal of the pulse have been proposed: improving the contrast of the pulse (using for instance saturable absorbers or high energy oscillator pulses) before injecting it in the laser regenerative amplifier, which is largely responsible for the ASE amplification [55], or using nonlinear optical processes, such as frequency doubling at the output of the laser (as employed for instance in [56]), and optical parametric CPA [57]. A very efficient method is based on the self-induced plasma shuttering or plasma mirror (PM) technique [58, 59]. The principle of the plasma mirror is the following: the laser pulse to be "cleaned" is focused with a proper choice of flux density onto a dielectric plate with an AR coating. The low intensity pedestal is almost completely transmitted, but as soon as the pulse intensity becomes high enough to ionize the dielectric, a plasma is formed and the following, more intense, part of the pulse is reflected at the critical density n_{c}. Using this technique, we had already demonstrated an improvement of the temporal contrast by a factor 100 for our 10 TW Ti-Sapphire laser system (600 mJ, 65 fs CPA system), leading to an overall contrast better than 10^8 and an energy transmission of 70% [60].

For routinely available peak intensities above $10^{18}\,\mathrm{W\,cm^{-2}}$, a 10^8 contrast is not sufficient to avoid the generation of a pre-plasma. As a consequence, we decided to use two consecutive reflections on a set of AR parallel dielectric plates [61], giving rise to a double plasma mirror. According to the performance of a single PM, we expected, in the optimum configuration, a gain of 10^4 and a transmission of 50%. The DPM was implemented before the optics that focus the beam onto the target, in a confocal set-up. The position of both PMs with respect to the focal spot determines the quality of the DPM in term of contrast, gain, and energy transmission as well as the quality of the focal spot. The laser beam is guided into the vacuum chamber housing the DPM, ideally placed between the compressor and the experimental chamber (see Fig. 10.3), after a compression of the laser pulse to 65 fs. A pair of retractable mirrors inside the DPM chamber direct the beam through the DPM; these can be removed to bypass the plasma mirrors. The laser beam is first focused by a dielectric $f = 1230\,\mathrm{mm}$ off-axis parabola between the two plasma mirrors.

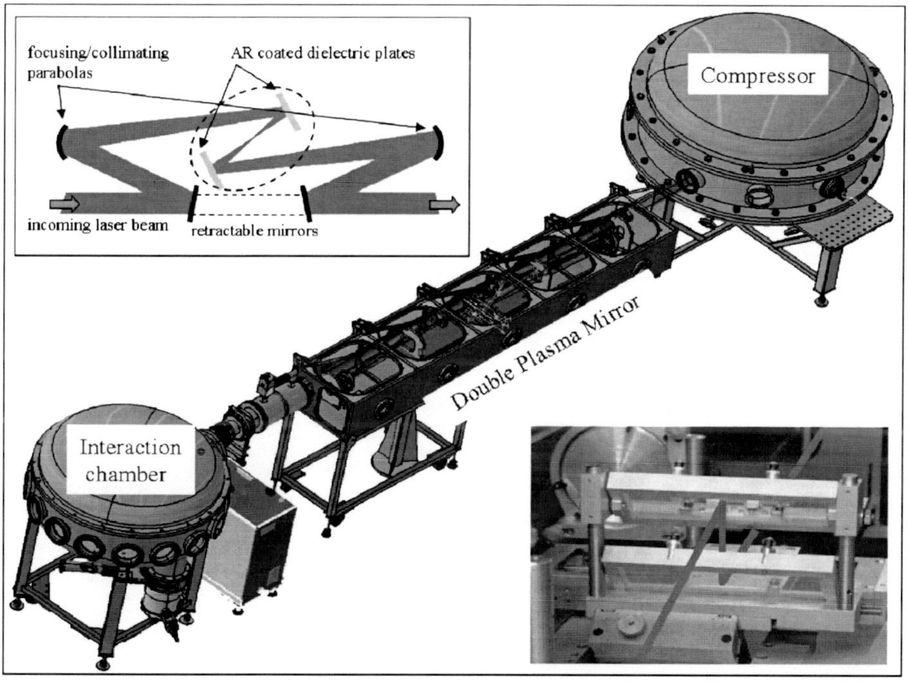

Fig. 10.3. Beam compressor, DPM, and experimental chambers layout. The *left* inset shows the beam path inside of the DPM, while the *right* inset shows the two AR dielectric plates (the *line* shows the beam path between them)

The position of the focal spot between the two PMs can be varied using an external motorized stage. After reflection on each PM, the beam is collimated by a second identical off-axis parabola and sent as a parallel beam in the experimental chamber. The entire setup is maintained under clean vacuum down to at least 10^{-6} mbar.

The laser beam is *s*-polarized with respect to each PM surface to maximize the PM reflectivity [60]. The distance between the PMs is about 4 cm. We used an anti-reflection coating, covering a spectral range from 750 to 850 nm, deposited on a dielectric bulk material(BK7; 150 × 40 × 20 mm). The DPM is automatically translated to offer a fresh surface after each shot and a drilled plate set between the two PMs avoids contamination of the mirror surfaces by debris, and the resulting reduction in the transmission coefficient. Our DPM allows about 1, 800 shots before replacement of both mirrors and can operate at around 1 Hz. We used a joule-meter to get the measure of the laser energy with or without the DPM, and find a transmission ratio of 50%. To measure the focal spot, we placed a 16 bit CCD camera just after an imaging relay system. The image of the focal spot is shown in Fig. 10.4. The width is identical with and without the DPM, suggesting that the DPM does not introduce any significant distortion of the beam wavefront. In our focusing conditions, the full width at half maximum is found to be approximately 8μm, giving a peak intensity of about $6 \cdot 10^{18}$ W cm^{-2}. We measured the temporal contrast with a high dynamic range (10^{12}) SEQUOIA third order cross-correlator (Amplitude Technologies) with a temporal resolution of 120 fs, placed just after an optical recovery system. The related contrast and focal spot are shown in Fig. 10.4. Each point of the contrast profile is an average

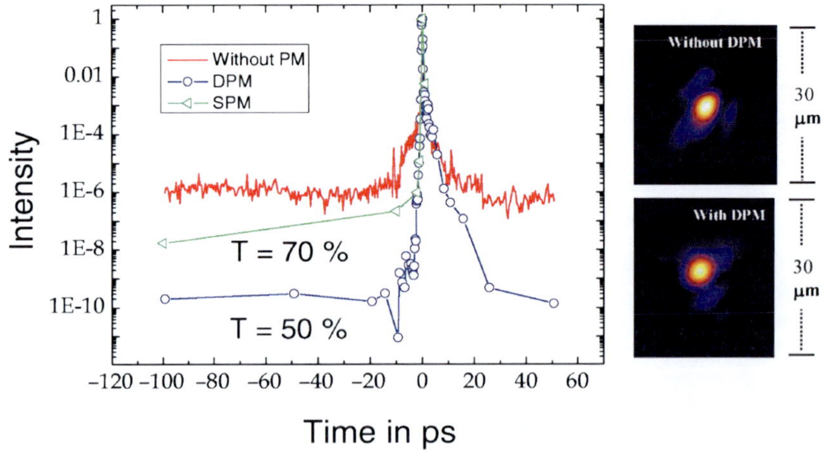

Fig. 10.4. Temporal profiles of the laser in the DPM configuration (*circles*), using a SPM (*triangles*) and without correction (*line*). On the *right*, the focal spots obtained without using the DPM (*top*) and with the DPM (*bottom*)

over three shots. For comparison, the results obtained with a single plasma mirror (SPM) (from [61]) and without any correction are also presented.

The contrast of the UHI10 laser is initially about 10^5 at 2 ps and 7×10^5 at long delays. With the DPM, we can see that the signal amplitude for long delays (ASE) starts from 10^{10} and grows by eight orders of magnitude in 2 ps (coherent contrast region). With the SPM, for the same delays we find respectively 10^8 and 10^6. It is noted that for both the transmission and for the contrast, the DPM behaves exactly like a "SPM to the power two" ($70\%^2 = 50\%$, $100^2 = 10^4$). Note that the temporal profile also exhibits a correction on the post-pulse side. This behavior, predicted by the modeling [60], is attributed to the plasma expansion producing a significant defocusing of the post-pulse after a few picoseconds.

The DPM was revealed to be a very robust device and is now routinely used for most of the experiments carried out on the UHI 10 TW laser chain involving ion acceleration as well as high order harmonics generation from solid targets [62].

10.4 Ion Acceleration at UHC

10.4.1 Acceleration Symmetry

As described earlier, one of the envisaged paths to increase the maximum laser-driven proton energy lies in reducing the target thickness. Taking advantage of our DPM performance, we carried out an experimental campaign to investigate proton and ion acceleration using mylar foils with a range of thicknesses spanning from about 100 µm to 800 nm. We used p-polarized pulses focused on targets with an off-axis $f = 300$ mm parabola with 45° incidence angle, resulting in an intensity of 10^{19} W cm^{-2} in low contrast conditions (LC) and 5×10^{18} W cm^{-2} in the ultra high contrast regime (UHC). We performed ion beam energy measurements using two similar Thomson parabola spectrometers simultaneously recording the BWD and FWD emission for each single shot at distances of 240 and 600 mm from target, respectively (inset in Fig. 10.5). The entrance pin-hole diameters were respectively 100 and 200 µm. Once dispersed by the magnetic and electric fields of the spectrometer, protons and ions were detected by a two stage 40 mm diameter micro channel plate (MCP) coupled to a phosphor screen. This was imaged onto a 12 bit CCD camera and the ion traces analyzed using simulated ion parabolas obtained through the commercial code SIMION [63]. The MCP efficiency for proton energies in our range of interest was assumed linear [64]. The spatial properties of the proton beams were characterized using a set of radio-chromic films (RCF) dosimetry media (Gafchromic HD-810) filtered by Al foils of properly chosen thickness in order to record proton energies ranging from 465 keV to 1.9 MeV.

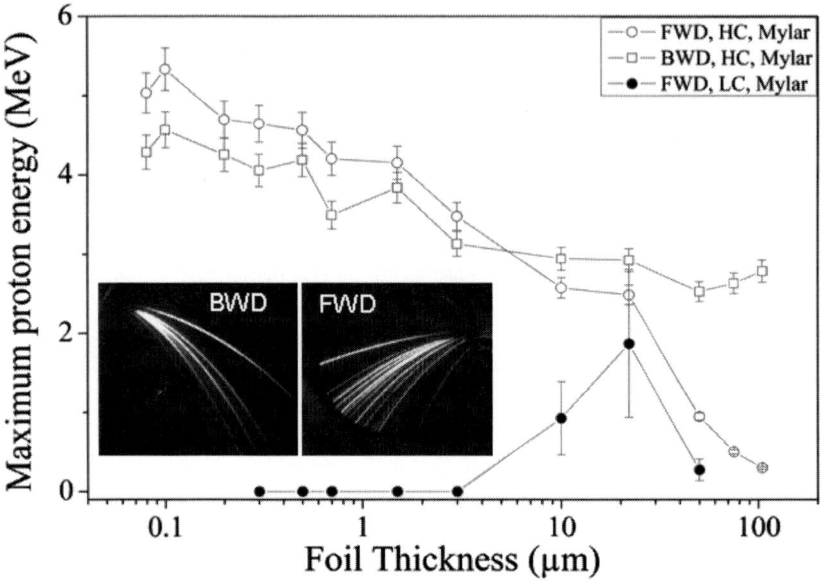

Fig. 10.5. Dependence of the FWD and BWD maximum proton energy on target thickness in low contrast and ultra high contrast conditions. The insert shows a typical output of the Thomson parabolas for the same shot on a sub-micrometer target. After protons, carbon ions result in the most intense signals

Although other ions, mainly carbon, have also been detected, here we restrict our analysis to protons only. Maximum detectable proton energies are reported in Fig. 10.5. Each point represents an average over at least three shots. Error bars correspond to the average standard deviation we found for three different test target thicknesses, repeating the same measure over around 30 shots. In the LC conditions, we detected no ion emission in the BWD direction. Considering the arguments in Sect. 10.2, this is not surprising. The pre-plasma produced by the pulse pedestal impedes any "TNSA-like" mechanism taking place for BWD emission and intrinsically imposes an optimum thickness for maximizing FWD emitted proton energies. As expected (Fig. 10.2II), for a mylar target the optimum thickness is about 20 μm. In the UHC regime, we observed an increase of the maximum proton energy by reducing the target thickness, thus confirming theoretical predictions and previously reported experimental results [35, 65, 66]. In more detail, at UHC the highest detected proton energy (at 0.1 μm) is more than twice that obtained with LC shots, although the UHC laser intensity is only half of that for LC. More surprisingly, for foil thicknesses lower than 10 μm, we observed quite similar maximum proton energies in both directions as though the TNSA mechanism applies for both sides of the target. Actually, for UHC shots, the ion density profile on the two sides of the target can be considered to be almost the same as the generation of a pre-plasma on the irradiated side,

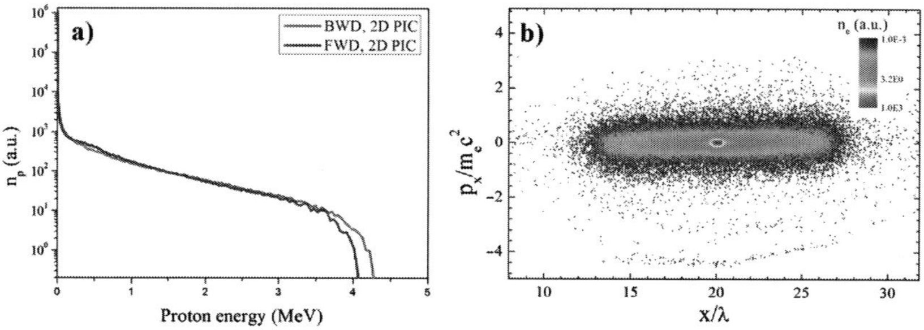

Fig. 10.6. 2D CALDER simulations for UHC experimental parameters; FWD and BWD proton energy distributions (**a**) and related electron phase space plot (**b**) at the laser peak

as well as the action of a shock wave propagating through the target, can be neglected. The electrostatic field created around the target by the hot electron cloud therefore has the same strength on both target sides. As a consequence, proton emissions with similar energies are expected in the FWD and BWD directions. This assumption is supported by the results of numerical simulations (detailed further below), showing a fairly symmetrical electron phase space at the laser peak and very similar peak proton energies (Fig. 10.6). Assuming similar acceleration properties for both directions implies that proton beams should show quite similar overall characteristics too. Indeed, the flux and divergence angle of protons emitted from both sides of the target are quite close to each other, especially for the highest recorded energies (see RCF recorded patterns for a 800 nm target in Fig. 10.7 (left)). We observed a spreading half-angle of 4° in the horizontal plane and 7° in the vertical one for both 3.2 MeV proton beams. In agreement with [67], the profiles of the proton beams show an elliptic shape whose bigger axis is oriented perpendicularly to the extended shape of the laser focal spot due to the 45° incidence angle. In the UHC regime, and especially for BWD emission, we observed an increasing carbon ion population upon reducing the target thickness. These carbon ions could be partly responsible [68] for the higher size of the BWD low-energy RCF pattern compared to the FWD one, observed in Fig. 10.7 (right).

The proton beam laminarity has been checked by imaging a 12.7 μm step grid placed parallel to the target onto RCF (Fig. 10.8). Using the line-outs of the grid on the RCF [69] and the profiles from the Thomson Parabola [70], we estimated an emittance of $0.06\pi \cdot \mathrm{mm} \cdot \mathrm{mrad}$ for the FWD beam and $0.12\pi \cdot \mathrm{mm} \cdot \mathrm{mrad}$ for the BWD.

As a whole, our set of measurements therefore shows that, contrary to what is usually observed under low contrast conditions [21, 71], BWD protons originating from the laser-irradiated surface have practically the same characteristics as the FWD protons from the target rear side, suggesting a "quasi-symmetrical" TNSA acceleration in the two directions.

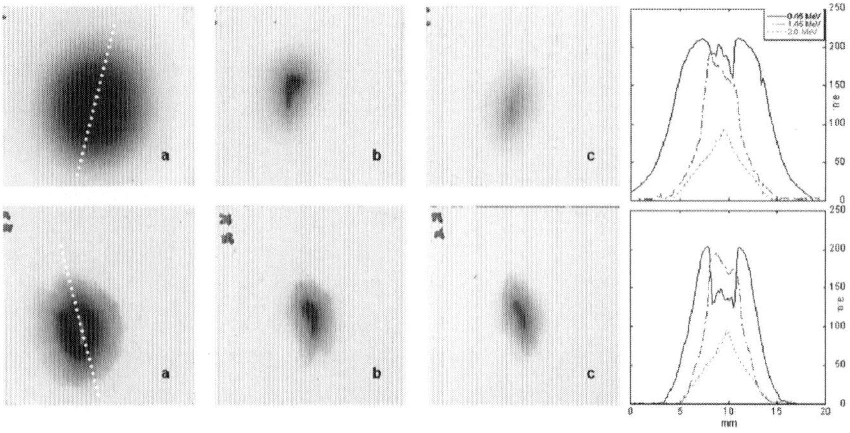

Fig. 10.7. (*left*) Radio-chromic film profiles in the BWD (*top*) and FWD (*bottom*) direction for the same shot. The recorded energies are 0.45 MeV (**a**), 1.45 MeV (**b**), and 2 MeV (**c**). Images are obtained adding an aluminum layer of 0.8, 25, and 40 μm, respectively, on the RCF. The *white dashed lines* show the direction of the line-outs shown to the right of the picture. Profile clipping at the center of the curves is due to the alteration of the outermost films by the high proton doses

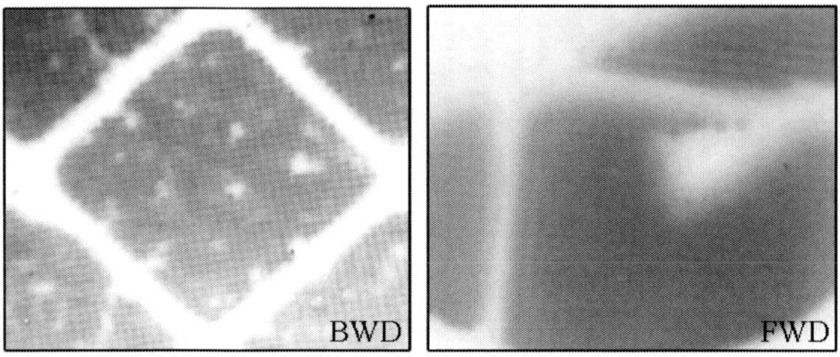

Fig. 10.8. Shadow of grid meshes (periods: 500 and 12.7 μm) made on radio-chromic film by the 0.45 MeV component of the BWD (*left*) and FWD (*right*) proton beam. In both images, the small step grid is clearly visible

Finally, experimental results have been compared with 2D (CALDER [72]) particle-in-cell simulations accounting for the UHC experimental parameters in terms of laser angle of incidence (45°), duration (65 fs), focal spot on the target (8 μm), intensity (3×10^{18} W cm^{-2}, slightly lower than the experimental value to take into account the lack of 3D effects in the code), and high target density (150 n_c). The simulations have been run over more than 560 fs, allowing for efficient energy exchange between heated electrons and ions. The computed peak proton energies in the FWD and BWD directions

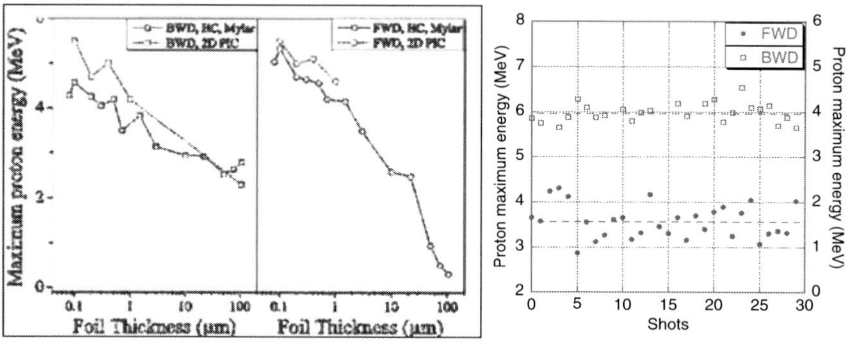

Fig. 10.9. Comparison of experimental data and 2D PIC simulations results for FWD and BWD maximum proton energy as a function of the target thickness (*left*), and peak proton energy shot-to-shot variation in UHC regime for FWD and BWD emission (*right*)

are plotted in Fig. 10.9. Simulations take the electron transport divergence through the target into account correctly [43] as is shown by the fairly good agreement observed over the entire explored range of thicknesses. FWD and BWD peak ion energies vary similarly both in experiments and simulations, showing a slight increase with decreasing target thickness. According to the code results, this observed thickness dependence on energy increase is mainly due to the increase of the electron cloud density around the target. Actually, both the calculated laser energy absorption ratios and electron temperatures show only a very weak increase as the targets become thinner.

10.4.2 Laser Energy Transfer

In small scale-length plasmas, the dominant heating mechanisms are essentially the $j \times B$ absorption and the resonance absorption. However, for a laser beam obliquely impinging on a very steep density gradient plasma, we expect the Brunel effect [39] to be mainly responsible for the energy transfer to the hot electron population. This mechanism cannot take place effectively for an s-polarized pulse or at normal incidence, where electrons are not dragged across the target surface, but along it. To corroborate this hypothesis, we experimentally checked the influence of the laser polarization on the maximum proton energy by inserting a zero-order half wave plate into the beam path between the DPM and the focusing parabola, and recorded the proton emission in the FWD and BWD directions for LC and UHC conditions. Data obtained for a 13 μm mylar target are presented in Fig. 10.10. For both emission directions, proton energies continuously decrease when the laser polarization is varied from p (0 wave plate angle) to s (45 angle). For an s-polarized electric field, the signal completely vanishes. It is worth noticing that under low contrast conditions, the peak proton energy for the s-polarized

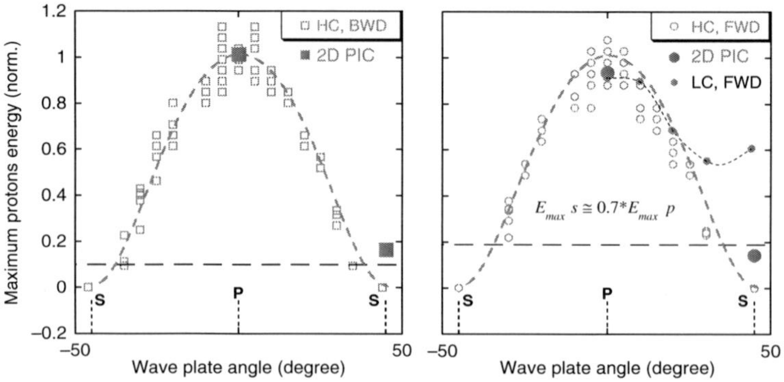

Fig. 10.10. Dependence of the maximum proton energy on the laser polarization for a 13 μm mylar target in UHC conditions in the BWD (*left*) and in the FWD (*right*) direction, and 2D PIC simulations results for p and s polarization. The *dashed line* represents $E_{\max} \propto \eta \times \sqrt{\cos(2\theta)^2}$

pulse was found to be only 35% weaker than for a p-polarized pulse, in fairly good agreement with previously reported, standard contrast experiments [73]. The maximum proton energy is expected to be proportional to the hot electron temperature. If we assume the Brunel effect to be the main energy transfer mechanism, the hot electron temperature is proportional to $\sqrt{I_{\mathrm{L}}}$ and the laser absorption for a laser beam impinging at 45° is, in the weakly relativistic regime [74], $\eta = (f/\pi a_0)[(1+f^2 a_0^2/2)^{1/2} - 1]$, where $f = 1+(1-2a_0/\pi)^{1/2}$ and a_0 is the standard dimensionless parameter ($a_0 = eE_{\mathrm{L}}/m_{\mathrm{e}}\omega_0 c$). The dependence of the maximum proton energy on the laser polarization can be then obtained replacing E_{L} in the above formula with the p-component of the laser intensity that is, $\cos(2\theta)$, where θ is the wave plate angle. The resulting expression $E_{\max} \propto \eta \times \sqrt{\cos(2\theta)^2}$ shows a fairly good agreement with experimental data, as shown in Fig. 10.10, suggesting that for short and high contrast laser pulses with intensities of about 10^{18} W cm^{-2}, the hot electron and proton energies are indeed mostly related to the p-component of the laser electric field through the Brunel effect.

10.5 An Analytical Model for Ion and Proton Emission at UHC

We would like to end our contribution by reporting some recently obtained results, which stress the two main practical advantages of the UHC regime discussed in Sect. 10.2: the possibility of using very thin targets and the improved shot-to-shot reproducibility. This latter point makes UHC interaction experiments a good benchmark to test simulation codes and/or analytical models.

In this framework, we have developed an analytical model based on a self-consistent solution of the Poisson equation using an adiabatic approximation for laser generated fast electrons [75]. This model, briefly outlined in the following, allows the determination of the optimal target thickness to optimize the maximum proton (and ion) energies, as well as the particle number as a function of given UHC laser pulse parameters.

At first, we extended the explored target thickness range down as far as 20 nm, observing a decrease in peak proton as well as carbon ion energy (Fig. 10.11). Because of a different experimental set-up, the laser intensity on target was slightly lower (10^{18} W cm^{-2}) and we used ultra-thin carbon foils for target thicknesses lower than 80 nm. Despite these differences, collected data seem to indicate that for our laser parameters, a 100 nm thickness actually maximizes the proton energy. In other words, with the UHC regime the optimal thickness actually shifts from 20 μm to 100 nm with a gain in peak energy higher than a factor 2. Moreover, we observed that for such a thickness, the accelerated particle number is optimized too.

This variation of ion energy and population as a function of the thickness is well described by our model, which uses a self-consistent solution of the Poisson equation for the electric field responsible for ion acceleration, and the equation of ion front motion describing the target expansion. A similar method was used in [51], although the above-mentioned system of equations was solved only numerically and for targets with just one kind of ion. Our model is more realistic than the solution of the Poisson equation at constant ion profile and solves the ion equation of motion in a determined field. Contrary to [76], fast electrons are described using an adiabatic instead of isothermal approximation because the main stage of ion acceleration in the expanding plasma usually begins after the end of the laser pulse. We used a 2D, two ion species PIC code [77] to get a detailed picture of the acceleration. The information we obtained concerning the symmetry of the plasma expansion, the shape of the accelerating electric field, and the duration of the acceleration process allowed the development of a 1D analytical model where the target is described by a heavy ion density distribution with a rectangular shape of varying width, whereas the hydrogen density distribution on the target surface is modeled by a δ function. The Poisson equation is of the following form:

$$\frac{\mathrm{d}^2\varphi}{\mathrm{d}\xi^2} = \left(1 + \frac{2}{3}\varphi\right)^{\frac{1}{2}} - \frac{N}{b}\theta(b-\xi) - \sigma\delta(\xi - b). \qquad (10.5)$$

Here n_{e_0} and T_{e_0} are the initial density and temperature of the fast electrons, $\varphi = \mid e \mid \phi/T_{\mathrm{e}_0}$ is the normalized potential, $\xi = z/r_{\mathrm{De}}$, $r_{\mathrm{De}}^2 = T_{\mathrm{e}_0}/4\pi e^2 n_{\mathrm{e}_0}$, $\sigma = Z_1 n_{i_1}\ell_{i_1}/n_{\mathrm{e}_0} r_{\mathrm{De}}$, $N = Z_2 n_{i_2}\ell_{i_2}/2n_{\mathrm{e}_0} r_{\mathrm{De}}$, θ is the unit step function, n_{i_j} and ℓ_{i_j} are, respectively, the ion density and layer thickness ($j = 1$ for hydrogen, 2 for heavy ions), and $b = b(t)/r_{\mathrm{De}}$ is the normalized distance of the ion front. The potential vanishes at infinity, which ensures the conservation of the electron number, and, due to the symmetry of the plasma slab, is zero at $\xi = 0$. The Poisson equation (10.5) can be solved analytically:

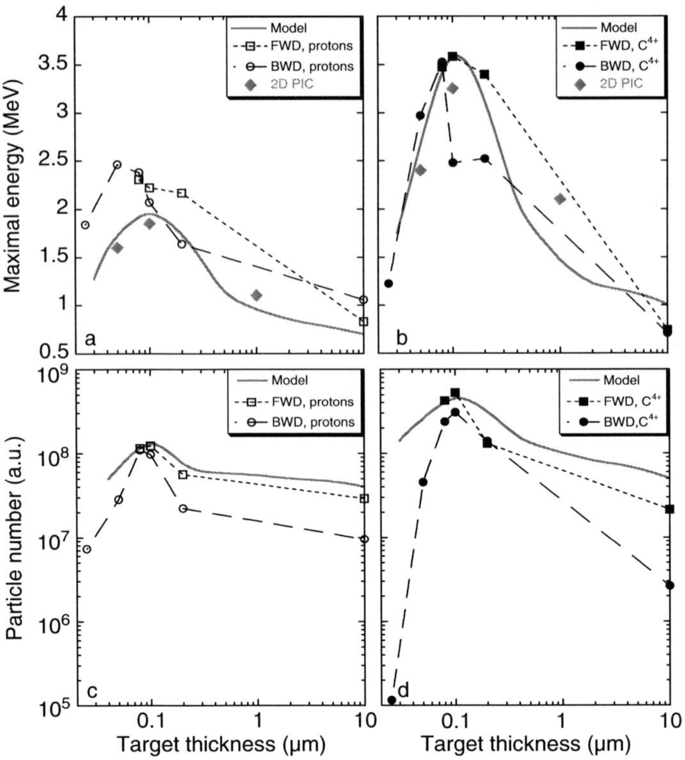

Fig. 10.11. Maximal experimental energies and number of particles for protons (**a**) and (**c**) and carbon ions (**b**) and (**d**) as a function of foil thickness. *Open* and *closed black circles* and *squares* are experimental data. The *solid lines* are the estimates from the analytical model. *Closed diamonds* are 2D PIC code results

$$\tilde{\varphi}(\xi) = \frac{1}{4 \times 3^4}\left[b - \xi + (4 \times 3^4 \tilde{\varphi}(b))^{1/4}\right]^4 \qquad \xi \geq b, \tag{10.6}$$

$$b - \xi(\tilde{\varphi}) = \int_{\tilde{\varphi}(b)}^{\tilde{\varphi}} \frac{\mathrm{d}\tilde{\varphi}}{\sqrt{\frac{8}{9}(\tilde{\varphi}^{3/2} - \tilde{\varphi}(0)^{3/2}) - \frac{4N}{3b}(\tilde{\varphi} - \tilde{\varphi}(0))}} \qquad 0 \leq \xi \leq b, \tag{10.7}$$

where $\tilde{\varphi} = (1 + (2/3)\varphi)$. It is possible to show that the electric fields before (+) and beside (−) the hydrogen layer can be expressed as

$$E^{\pm}(b) \approx \left(\frac{E_{\delta}^{\pm}}{\sqrt{1 + (\ell_{i_2}/r_{\mathrm{De}})^2}} + \frac{E_{\infty}^{\pm}}{\sqrt{1 + (r_{\mathrm{De}}/\ell_{i_2})^2}}\right) \times \left(\frac{1}{1 + (b/r_{\mathrm{De}})^{3/2}}\right), \tag{10.8}$$

where $E_{\delta}^{\pm} = 2\pi e(Z_2 n_{i_2} \ell_{i_2} \pm Z_1 n_{i_1} \ell_{i_1})$ and $E_{\infty}^{\pm} = \sqrt{4\pi Z_2 n_{i_2} T_{\mathrm{e}0}}(\sqrt{2}/3^{3/4})$ $(Z_2 n_{i_2}/n_{\mathrm{e}0}) \pm 2\pi e Z_1 n_{i_1} \ell_{i_1}$, respectively, for very thin ($b \ll 1$) and thick ($b \gg 1$) targets. Consequently, the equation of motion of heavy and light ions are

$$m_{i_{1,2}} \frac{\partial^2 b_{1,2}}{\partial t^2} = Z_{1,2} e \left(\frac{E_\delta^\pm}{\sqrt{1 + (\ell_{i_2}/r_{\mathrm{De}})^2}} \frac{1}{1 + (b_{1,2}/r_{\mathrm{De}})^{3/2}} + \frac{E_\infty^\pm}{\sqrt{1 + (r_{\mathrm{De}}/\ell_{i_2})^2}} \frac{1}{1 + (b_{1,2}/r_{\mathrm{De}})^{3/2}} \right), \quad (10.9)$$

where $b_{1,2}$ are the coordinates of light and heavy ion fronts. Using the energy conservation law for this last expression and estimating the fast electron temperature through 2D PIC code simulations allows us to obtain an expression for the variation of the maximum energy for protons and heavy ions, and to estimate the particle number as a function of the target thickness. Figure 10.11 shows the fair agreement between this model, the 2D PIC code results and the set of data we collected in Saclay using ultra-thin carbon as well as mylar targets.

References

1. S.J. Gitomer, R.D. Jones, F. Begay, A.W. Ehler, J.F. Kephart, R. Kristal, Phys. Fluid **29**, 2679–2688 (1986)
2. A.P. Fews, P.A. Norreys, F.N. Beg, A.R. Bell, A.E. Dangor, C.N. Danson, P. Lee, S.J. Rose, Phys. Rev. Lett. **73**, 1801–1804 (1994)
3. D. Strickland, G. Mourou, Opt. Commun. **56**, 219–221 (1985)
4. P. Maine, D. Strickland, P. Bado, M. Pessot, G. Mourou, IEEE J. Quant. Electron. **24**, 398–403 (1988)
5. S.P.D. Mangles, C.D. Murphy, Z. Najmudin, A.G.R. Thomas, J.L. Collier, A.E. Dangor, E.J. Divall, P.S. Foster, J.G. Gallacher, C.J. Hooker, D.A. Jaroszynski, A.J. Langley, W.B. Mori, P.A. Norreys, F.S. Tsung, R. Viskup, B.R. Walton, K. Krushelnick, Nature **431**, 535–538 (2004)
6. C.G.R. Geddes, C. Toth, J. van Tilborg, E. Esarey, C.B. Schroeder, D. Bruhwiler, C. Nieter, J. Cary, W.P. Leemans, Nature **431**, 538–541 (2004)
7. J. Faure, Y. Glinec, A. Pukhov, S. Kiselev, S. Gordienko, E. Lefebvre, J.-P. Rousseau, F. Burgy, V. Malka, Nature **431**, 541–544 (2004)
8. W.P. Leemans, B. Nagler, A.J. Gonsalves, C. Tóth, K. Nakamura, C.G.R. Geddes, E. Esarey, C.B. Schroeder, S.M. Hooker, Nat. Phys. **2**, 696–699 (2006)
9. E. Esarey, R.F. Hubbard, W.P. Leemans, A. Ting, P. Sprangle, Phys. Rev. Lett. **79**, 2682–2685 (1997)
10. J. Faure, C. Rechatin, A. Norlin, A. Lifschitz, Y. Glinec, V. Malka, Nature **444**, 737–739 (2006)
11. M. Borghesi, et al., Phys. Rev. Lett. **94**(19), 195003 (2005)
12. M. Tabak, et al., Phys. Plasma **1**, 1626–1634 (1994)
13. K.W.D. Ledingham, et al., Science **300**, 1107–1111 (2003)
14. M.I.K. Santala, et al., Appl. Phys. Lett. **78**, 19–21 (2001)
15. S.S. Bulanov, et al., Med. Phys. **35**(5), 1770–1776 (2008)
16. M. Borghesi, J. Fuchs, S.V. Bulanov, A.J. Mackinnon, P.K. Patel, M. Roth, Fusion Sci. Technol. **49**, 412–439 (2006)
17. V. Malka, J. Faure, Y.A. Gauduel, E. Lefebvre, A. Rousse, K.T. Phuoc, Nat. Phys. **4**, 447–453 (2008)

18. S.C. Wilks, et al., Phys. Plasma **8**(2), 542–549 (2001)
19. R.A. Snavely, et al., Phys. Rev. Lett. **85**, 2945–2948 (2000)
20. E.L. Clark, et al., Phys. Rev. Lett. **84**, 670–673 (2000)
21. A. Maksimchuk, S. Gu, K. Flippo, D. Umstadter, V. Yu. Bychenkov, Phys. Rev. Lett. **84**(18), 4108–4111 (2000)
22. L.V. Pariiskaya A.V. Gurevich, L.P. Pitaevskii, Sov. Phys. JETP **22**, 449 (1966)
23. J. Fuchs, et al., Phys. Rev. Lett. **94**(4), 045004 (2005)
24. C.K. Li, F.H. Séguin, J.R. Rygg, J.A. Frenje, M. Manuel, R.D. Petrasso, R. Betti, J. Delettrez, J.P. Knauer, F. Marshall, D.D. Meyerhofer, D. Shvarts, V.A. Smalyuk, C. Stoeckl, O.L. Landen, R.P.J. Town, C.A. Back, J.D. Kilkenny, Phys. Rev. Lett. **100**(22), 225001 (2008)
25. A.J. MacKinnon, P.K. Patel, M. Borghesi, R.C. Clarke, R.R. Freeman, H. Habara, S.P. Hatchett, D. Hey, D.G. Hicks, S. Kar, M.H. Key, J.A. King, K. Lancaster, D. Neely, A. Nikkro, P.A. Norreys, M.M. Notley, T.W. Phillips, L. Romagnani, R.A. Snavely, R.B. Stephens, R.P.J. Town, Phys. Rev. Lett. **97**(4), 045001 (2006)
26. U. Amaldi, Nucl. Phys. A **654**, 375–399 (1999)
27. J. Fuchs, et al., Nat. Phys. **2**, 48–54 (2006)
28. P. Mora, Phys. Rev. Lett. **90**(18), 185002 (2003)
29. L. Robson, P.T. Simpson, R.J. Clarke, K.W.D. Ledingham, F. Lindau, O. Lundh, T. McCanny, P. Mora, D. Neely, C.-G. Wahlström, M. Zepf, P. McKenna, Nat. Phys. **3**, 58–62 (2007)
30. P. Mora, Phys. Rev. E **72**, 056401 (2005)
31. A.P.L. Robinson, M. Zepf, S. Kar, R.G. Evans, C. Bellei, New J. Phys. **10**(1), 013021 (2008)
32. T.V. Liseikina, A. Macchi, Appl. Phys. Lett. **91**(17), 171502 (2007)
33. A. Macchi, F. Cattani, T.V. Liseykina, F. Cornolti, Phys. Rev. Lett. **94**(16), 165003 (2005)
34. Y. Sentoku, et al., Phys. Plasma **10**, 2009 (2003)
35. A.J. Mackinnon, Y. Sentoku, P.K. Patel, D.W. Price, S. Hatchett, M.H. Key, C. Andersen, R. Snavely, R.R. Freeman, Phys. Rev. Lett. **88**(21), 215006 (2002)
36. I. Spencer, et al., Phys. Rev. E **67**(4), 046402 (2003)
37. Paul Gibbon, Phys. Rev. Lett. **73**(5), 664–667 (1994)
38. P. Gibbon, A.R. Bell, Phys. Rev. Lett. **68**(10), 1535–1538 (1992)
39. F. Brunel, Phys. Rev. Lett. **59**(1), 52–55 (1987)
40. W.L. Kruer (ed.), The Physics of Laser Plasma Interactions, vol 73 (1988) Perseus Books, U.S.
41. G. Malka, J.L. Miquel, Phys. Rev. Lett. **77**(1), 75–78 (1996)
42. H. Popescu, S.D. Baton, F. Amiranoff, C. Rousseaux, M.R. Le Gloahec, J.J. Santos, L. Gremillet, M. Koenig, E. Martinolli, T. Hall, J.C. Adam, A. Heron, D. Batani, Phys. Plasma **12**(6), 063106 (2005)
43. M. Kaluza, et al., Phys. Rev. Lett. **93**(4), 045003 (2004)
44. J.J. Honrubia, M. Kaluza, J. Schreiber, G.D. Tsakiris, J. Meyer-Ter-Vehn, Phys. Plasma **12**(5), 052708 (2005)
45. H.J. Lee, K.H. Pae, H. Suk, S.J. Hahn, Phys. Plasma **11**(4), 1726–1729 (2004)
46. H. Habara, R. Kodama, Y. Sentoku, N. Izumi, Y. Kitagawa, K.A. Tanaka, K. Mima, T. Yamanaka, Phys. Rev. E, **69**(3), 036407 (2004)
47. S.C. Wilks, W.L. Kruer, M. Tabak, A.B. Langdon, Phys. Rev. Lett. **69**(9), 1383–1386 (1992)

48. J.T. Seo, S.H. Yoo, S.J. Hahn, J. Phys. Soc. Jpn. **76**(11), 114501 (2007)
49. A.J. Mackinnon, M. Borghesi, S. Hatchett, M.H. Key, P.K. Patel, H. Campbell, A. Schiavi, R. Snavely, S.C. Wilks, O. Willi, Phys. Rev. Lett. **86**(9), 1769–1772 (2001)
50. J. Fuchs, C.A. Cecchetti, M. Borghesi, T. Grismayer, E. d'Humières, P. Antici, S. Atzeni, P. Mora, A. Pipahl, L. Romagnani, A. Schiavi, Y. Sentoku, T. Toncian, P. Audebert, O. Willi, Phys. Rev. Lett. **99**(1), 015002 (2007)
51. T. Grismayer, P. Mora, Phys. Plasma **13**(3), 032103 (2006)
52. A.A. Andreev, R. Sonobe, S. Kawata, S. Miyazaki, K. Sakai, K. Miyauchi, T. Kikuchi, K. Platonov, K. Nemoto, Plasma Phys. Contr. Fusion **48**, 1605–1619 (2006)
53. A. Lévy, et al., Opt. Lett. **32**, 310 (2007)
54. T. Ceccotti, et al., Phys. Rev. Lett. **99**(18), 185002 (2007)
55. J. Itatani, J. Faure, M. Nantel, G. Mourou, S. Watanabe, Opt. Commun. **148**, 70–74 (1998)
56. U. Teubner, G. Pretzler, T. Schlegel, K. Eidmann, E. Förster, K. Witte, Phys. Rev. A **67**(1), 013816 (2003)
57. R. Butkus, R. Danielius, A. Dubietis, A. Piskarskas, A. Stabinis, Appl. Phys. B Laser Optic. **79**, 693–700 (2004)
58. H.C. Kapteyn, M.M. Murnane, A. Szoke, R.W. Falcone, Optic. Lett. **16**, 490–492 (1991)
59. D.M. Gold, H. Nathel, P.R. Bolton, W.E. White, L.D. Van Woerkom, SPIE **1413**, 41 (1991)
60. G. Doumy, F. Quéré, O. Gobert, M. Perdrix, Ph. Martin, P. Audebert, J.C. Gauthier, J.-P. Geindre, T. Wittmann, Phys. Rev. E **69**(2), 026402 (2004)
61. P. Monot, et al., Optic. Lett. **29**, 893 (2004)
62. C. Thaury, F. Quéré, J.-P. Geindre, A. Levy, T. Ceccotti, P. Monot, M. Bougeard, F. Réau, P. D'Oliveira, P. Audebert, R. Marjoribanks, P. Martin, Nat. Phys. **3**, 424–429 (2007)
63. Scientific Instrument Services, SIMION, www.simion.com.
64. H. Schwoerer, et al., Nature **439**, 445–448 (2006)
65. D. Neely, et al., Appl. Phys. Lett. **89**(2), 021502 (2006)
66. P. Antici, et al., Phys. Plasma **14**, 701 (2007)
67. J. Fuchs, et al., Phys. Rev. Lett. **91**(25), 255002 (2003)
68. E. Brambrink, J. Schreiber, T. Schlegel, P. Audebert, J. Cobble, J. Fuchs, M. Hegelich, M. Roth, Phys. Rev. Lett. **96**(15), 154801 (2006)
69. M. Borghesi, et al., Phys. Rev. Lett. **92**(5), 055003 (2004)
70. Y. Oishi, et al., Phys. Plasma **12**(7), 073102 (2005)
71. P. McKenna, et al., Phys. Rev. E **70**(3), 036405 (2004)
72. E. Lefebvre, et al., Nucl. Fusion **43**, 629 (2003)
73. A. Fukumi, et al., Phys. Plasma **12**, 701 (2005)
74. P. Gibbon, *Short Pulse Interactions With Matter* (Imperial College Press, London, 2005)
75. A. Andreev et al, Phys.Rev. Lett. **101**, 155002 (2008)
76. B.J. Albright, et al., Phys. Rev. Lett. **97**(11), 115002 (2006)
77. A.J. Kemp, H. Ruhl, Phys. Plasma **12**(3), 033105 (2005)

Index